TURBO MACHINES

TURBO MACHINES

By

Vijeta Kumari

2016

SBS Publishers & Distributors Pvt. Ltd.

New Delhi

ISBN 13 : 9789380090863

First Published in 2016

Published by:

SBS PUBLISHERS & DISTRIBUTORS PVT. LTD.

2/9, Ground Floor, Ansari Road, Darya Ganj,

New Delhi - 110002,

INDIA

Tel: 0091.11.23289119 / 41563911

Email: mail@sbspublishers.com

www.sbspublishers.com

Preface

A turbo machine is a device in which energy transfer occurs between a flowing fluid and rotating element due to dynamic action. This results in change of pressure and momentum of the fluid. The word turbo implies a spinning action is involved. In turbomachinery a blade or row of blades rotates and imparts or extracts energy to or from the fluid. Work is generated or extracted by means of enthalpy changes in the working fluid. In general, two kinds of turbomachines are encountered in practice. These are open and closed turbomachines. Open machines such as propellers, windmills, and unshrouded fans act on an infinite extent of fluid, whereas, closed machines operate on a finite quantity of fluid as it passes through housing or casing. We will examine only turbomachines of the closed type. We also categorize turbomachines according to the type of flow. When the flow is parallel to the axis of rotation, we denote this type of machine as axial flow, and when flow is perpendicular to the axis of rotation, we denote this machine as radial flow. Finally, when both radial and axial flow velocity components are present, the machine is denoted as mixed flow. Turbomachines may be further classified into two additional categories: those that absorb energy to increase the fluid pressure, i.e. pumps, fans, and compressors, and those that produce energy such as turbines by expanding to lower pressures. Of particular interest are applications which contain pumps, fans, compressors, and turbines. These components are essential in almost all mechanical equipment systems such as power and refrigeration cycles. We will examine each of these components in detail, and address a number of operating issues in systems when more than one component exists.

Editor

Contents

Chapter 1

ON THE INSTABILITY THRESHOLD OF JOURNAL BEARING SUPPORTED ROTORS

Ricardo Ugliara Mendes and Katia Lucchesi Cavalca

Laboratory of Rotating Machinery, Department of Mechanical Design, Faculty of Mechanical Engineering, University of Campinas, 200 Mendeleev Street, 13083-970 Campinas, SP, Brazil

ABSTRACT

Journal bearing supported rotors present two kinds of self-excited vibrations: oil-whirl and oil-whip. The first one is commonly masked by the rotor unbalance, hence being rarely associated with instability problems. Oil-whip is a severe vibration which occurs when the oil-whirl frequency coincides with the first flexural natural frequency of the shaft. In many cases, oil-whip is the only fluid-induced instability considered during the design stage; however, experimental evidences have shown that the instability threshold may occur much sooner, demanding a better comprehension of the instability mechanism. In this context, numerical simulations were made in order to improve the identification of the instability threshold for two test rig configurations: one on which the instability occurs on the oil-whip frequency, and another which became unstable before this threshold. Therefore, the main contribution of this paper is to present an investigation of two different thresholds of fluid-induced instabilities and their detectability on design stage simulations based on rotordynamic analysis using linear speed dependent coefficients for the bearings.

INTRODUCTION

Rotating machinery supported by journal bearings present two kinds of self-excited vibrations, namely, oil-whirl and oil-whip, which can drive the system to an unstable condition.

The understanding of the stability threshold phenomena and how they influence the dynamic response of the system has been a topic of great concern since the first studies of the lubrication theory, dating from the 1880s with the works of Petrov [1–4], Tower [5, 6], and Reynolds [7].

Oil-whirl is a subsynchronous vibration, occurring near half the rotational speed of the shaft: 43% to 49% according to Oliver [8]. It is usually masked by rotor unbalance and it can usually be seen only in the bearings. Oil-whip is a severe vibration that occurs once the oil-whirl coincides with the first flexural natural frequency of the shaft, hence, near twice this natural frequency. When operating under this condition, the vibration amplitudes in the bearings are limited by the bearings clearance; however, the shaft vibration can be very high once the shaft vibrates in a resonant condition—its natural frequency. The oil-whip instability was first described in the works of Newkirk [9, 10], and a more extensive description of these self-excited vibrations was made in the surveys of Muszynska [11, 12], who also evaluated stability regions using simple models. According to Muszynska [11], often the term "unstable," when talking about oil-whirl and oil-whip, is rather close to the terms "undesired" or "unacceptable," once that both phenomena present a periodic pattern, which, however, inserts a subharmonic component as significant as the harmonic component due to unbalance, in this case, designated by instability. In its purest sense "instability" refers to vibrations growing without bound. In this case, a limit cycle subsynchronous vibration may have large orbits, limited by the bearing damping. However, in order to maintain the term usually used in the literature, the instability threshold refers here to the point when the oil-whirl phenomenon begins.

Therefore, the contribution of this work is the stability analysis process itself, that is, how the conventional rotordynamic analysis can lead to the understanding of the instability threshold in different rotor-bearings systems and how the information of the fluid frequencies can help in this matter, once these frequencies are not usually taken into account in the classical analysis.

Aiming at a better estimation of a rotor critical speed, Stodola [13] and Hummel [14] considered the flexibility of the bearing oil film, introducing the idea of representing its dynamic properties by means of springs and dampers. This idea was explored in a more formal way in the work of Lund [15], where equivalent linear coefficients of stiffness and damping were calculated through the simplified solution of the Reynolds equation. Recently, a finite-difference method was used to solve the Reynolds equation in the work of Machado and Cavalca [16], and a finite-volume method was used in Machado [17], both intending to calculate more accurate equivalent coefficients, based on the previous work of Lund.

During the development of the equivalent linear coefficients, other researches were made towards nonlinear methods; for example, Capone [18, 19] presented a nonlinear solution for the Reynolds equation, obtaining a nonlinear equation for the force exerted by the journal bearings on the shaft.

In this context, many authors [20–22] used finite element models (FEM) of shafts together with journal bearings models to increase the understanding of complex rotors. Lima [23] made simulations with Capone's model and extended it to tilting-pad bearings. Castro et al. [24] used the same bearing model with a FEM model based in the work of Nelson and McVaugh [25] to study the stability threshold of horizontal and vertical machines, achieving good agreement with experimental results.

In the existing literature, oil-whirl is commonly referred to as a weak self-excited vibration, nevertheless, treated as instability. The first part of this statement relies on the fact that oil-whirl effect is not observable in the rotor, once its vibration is dominated by the unbalance forces. The vibration amplitude of the shaft inside the bearings is a strong function of the eccentricity of the shaft within the bearing clearance, that is, the bearing eccentricity. An important concept regarding this matter is the eccentricity ratio, which is a dimensionless quantity given by the bearing eccentricity divided by the bearing radial clearance. Concerning vertical rotors or lightly loaded horizontal rotors (with very low eccentricity ratio) the oil-whirl vibration in the bearing is limited by the fluid forces and the bearing clearance; this occurs because, as discussed later, in these rotors the cross-coupled coefficients of stiffness increase drastically as the rotor accelerates, overcoming the dissipative forces of the damping coefficients.

In some cases, this vibration due to oil-whirl may cause heating and wear in the bearing, justifying the use of the term instability. In such light rotors, it is also possible that the oil-whirl component becomes very small or vanishes during the rotor crossover through the critical speed, giving place to the rise of the synchronous component due to unbalance [11]. For horizontal rotors with higher eccentricity ratio, the subsynchronous oil-whirl component may be present in the bearings but usually with small amplitudes.

Therefore, when talking about stability threshold on horizontal rotors, oil-whip is the phenomenon usually considered. However, experimental evidences have shown that, in some cases, the instability threshold may occur sooner than the expected oil-whip frequency. Such observations were made in a test rig at the Laboratory of Rotating Machinery in the University of Campinas, when simulating a procedure for stability identification of journal bearing supported machines.

Thereby, the correct identification of the instability threshold is mandatory to ensure the expected performance in a secure operating condition, still during the design stage, remarkably in the case of high speed flexible rotors (e.g., multistage rotors present in power plants). The previous knowledge of the instability threshold may be also useful to the development of control systems to stabilize the fluid-induced instabilities, allowing the rotors to safely operate in higher rotational speeds. Siqueira et al. [26] developed a linear parameter varying (LPV) control, which takes into account the parameter dependency of the system (in this case, the rotational speed), using the H-infinite technique. A significant vibration amplitude reduction during the crossing of the critical speed in a run-up simulation was obtained using a magnetic actuator. The authors stated that an important factor in the control design was a representative model; which they obtained using the linear coefficients approach. Riemann et al. [27] took a step further and managed to control the oil-whip instability using a μ-synthesis control, extending the rotor operational range from 42 Hz (oil-whip threshold) up to 60 Hz in an experimental test rig; a magnetic actuator was used as well. Tuma et al. [28] extended the instability threshold of a rigid rotor from 3400 rpm to 7300 rpm using two piezoactuators at the bearing bushing and a simple proportional controller.

In this paper, two test rig configurations—hence, with different dynamic behaviors—were simulated showing the instability present in each case. The models consist of a shaft FEM with linear coefficients of stiffness and damping representing the journal bearings. The different conditions of instability are compared through Campbell diagrams, orbits, and run-down simulations. Experimental evidences of oil-whirl and oil-whip are vastly discussed in the literature, mainly by the works of Muszynska [11, 12]. The experimental measurements which motivated this work are also presented and compared with the simulated results.

Therefore, the investigation of the fluid-induced instability in different conditions (here obtained through different test rig configurations) and its detectability in the design stage are the main contribution of this paper.

MATERIALS AND METHODS

Two test rig configurations were simulated to compare the different conditions of instability; the finite element models (FEM) used are presented in Figure 1, and details of both configurations are found in Table 1.

Table 1. Details concerning both testing configurations.

Shaft material	Steel SAE 1030
Shaft diameter	12 mm
Shaft length	800 mm (A)–820 mm (B)
Disc material	Steel SAE 1020
Disc diameter	95 mm
Disc width	47.5 mm
Shaft journal material	Steel SAE 1020
Shaft journal diameter	40 mm
Shaft journal width	80 mm
Coupling coefficients	$K_{yy} = K_{zz} = 1e5$ N/m and $K_{yz} = K_{zy} = 0$ N/m
Ball bearing coefficients	$K_{yy} = K_{zz} = 1e10$ N/m and $K_{yz} = K_{zy} = 0$ N/m
Journal bearings radial clearance	90 μm
Journal bearings diameter	30 mm
Journal bearings width	20 mm
Lubricant	ISO VG 32 oil

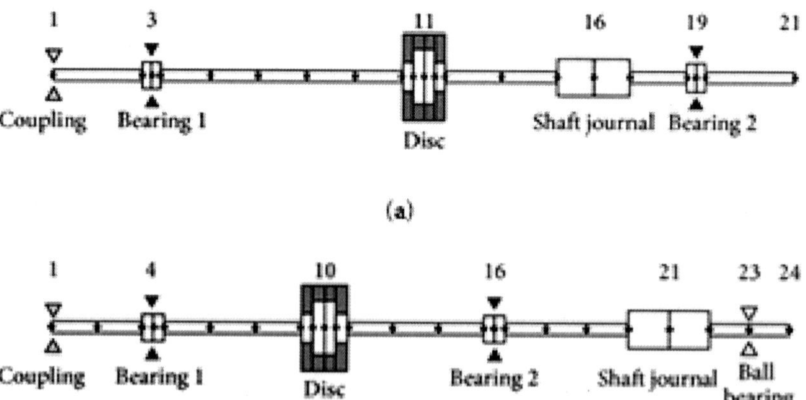

Figure 1. Finite element models: (a) configuration A; (b) configuration B.

Configuration A (Figure 1(a)) is the same from the works of Castro et al. [24] and Mendes et al. [29]. It contains a shaft supported by two hydrodynamic bearings symmetrically placed 600 mm from each other (nodes 3 and 19). It also contains a disc placed at the shaft mid span (nodes 9 to 13) and a journal at 62 mm of the first bearing (nodes 15 to 17). The connection between the end of the shaft and the WEG motor is made through a coupling (node 1) modeled by constant linear stiffness coefficients.

Configuration B (Figure 1(b)) consists of a shaft supported by three bearings. Two journal bearings are placed 350 mm from each other (nodes 4 and 16) and 103 mm from the right end of the shaft, where there is the motor coupling (node 1). The third bearing is a ball bearing at 305 mm from the second journal bearing (modeled by constant linear stiffness coefficients at node 23). In this configuration, the disc lays between the journal bearings (nodes 8 to 12) and the shaft journal at 105 mm before the ball bearing (nodes 20 to 22).

As it can be seen in Figure 1, the FEM model of configuration A was discretized in 21 nodes; while configuration B was discretized in 24 nodes. The position of the journal bearings is marked by black triangles, the ball bearing by a white triangle, and the coupling by a grey triangle.

The FEM matrixes used in the flexible rotor models were the classic Euler-Bernoulli beam matrixes for rotating shafts (i.e., including gyroscopic effect), as presented in Nelson and McVaugh [25] and here reproduced in the appendix. The equation of motion is given in (1), where is the rotational speed, is the mass matrix, is the gyroscopic matrix, is the stiffness matrix, and is the damping matrix (proportional to by a factor of according to Santana et al. [30]). Besides, the matrix is obtained from the sum of two matrices (2): the rotational mass matrix and the translational mass matrix . is the external forces vector, in this case, due to an unbalanced mass; is the generalized coordinate vector, with two translational and two rotational degrees of freedom (d.o.f.) for each node:

The journal bearings are modeled by equivalent coefficients of stiffness and damping (Figure 2), added to the corresponding d.o.f. at the bearings location in the stiffness and damping matrixes of the system, and , respectively. The speed dependent coefficients are evaluated from Machado [16, 17, 31] by the solution of the Reynolds equation (3) using finite volume method:

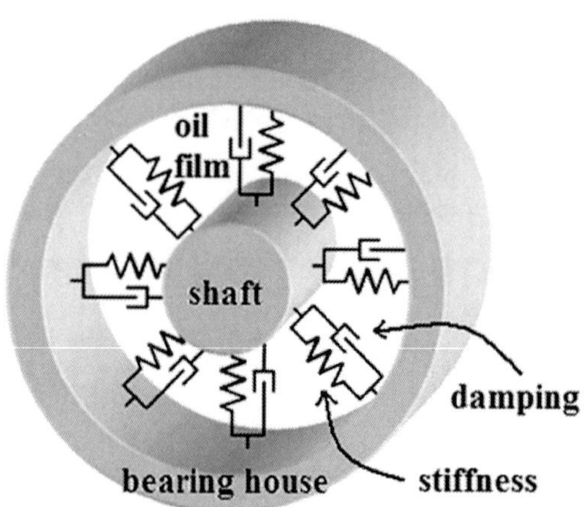

Figure 2. Model of the journal bearings.

In (3), is axial coordinate, is the circumferential coordinate, is the viscosity of the fluid, is the fluid film thickness, is the pressure, and is the time. From the fluid film thickness, both the linear coefficients (damping and stiffness) are calculated through the perturbation method. The pressure distribution is evaluated from Reynolds equation depending on the fluid film thickness expression. The reaction forces of the bearing can be obtained from the integration of pressure distribution. These forces are functions of the displacements and the instantaneous journal center velocities in direction and . Initially, the equilibrium position (,) of the shaft inside the bearing is found depending on the rotational speed. In this step, the term is not considered. Afterwards, the complete Reynolds equation (3) is solved again considering small deviations around the equilibrium position, and . At this point, a Taylor series expansion yields to (4), where the coefficients are partial derivatives evaluated at the equilibrium position (,) as given by (5):

Lund [15] pointed out that the solution of the Reynolds equation should be solved simultaneously with the dynamic equation of the rotor; still, for practical purposes it can be assumed that the vibration amplitude is small enough to allow substituting the fluid film forces by their gradients around the steady-state eccentricity for a given rotational speed; that is, the forces become proportional to the displacement (stiffness coefficients) and velocity (damping coefficients). Lund [32] also states that, although the coefficients definition is only valid for infinitesimal amplitudes, comparisons with exact solutions have shown that the coefficients approach may be applied for amplitudes as large as nearly forty percent of the bearing clearance. It is also recommended that the eccentricity ratio lies below 0.6-0.7.

In this work, the simulations performed presented much lower amplitude levels than the limit stated by Lund [32], except when the stability limit is reached. It is remarked here that the interest of the analysis performed is in finding out the instability threshold of the system and not the rotor behavior once it is unstable.

The coefficients behavior dependence on the rotational speed is shown in Figure 3 (stiffness coefficients) and Figure 4 (damping coefficients). The eccentricities of the journal bearings used in both configurations are presented in Figure 5, which shows that below 10 Hz, the bearings operate with eccentricity ratio less than 0.6, justifying the linearized coefficients behavior in this range.

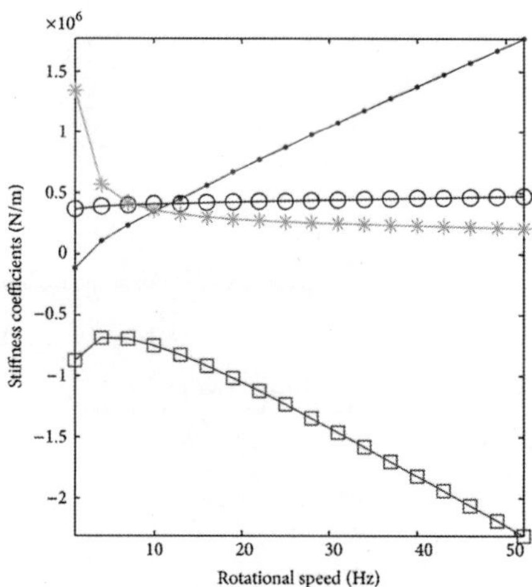

Figure 3. Equivalent coefficients of stiffness for the journal bearings: (a) bearing 1 in configuration A; (b) bearing 2 in configuration A; (c) bearing 1 in configuration B; (d) bearing 2 in configuration B.

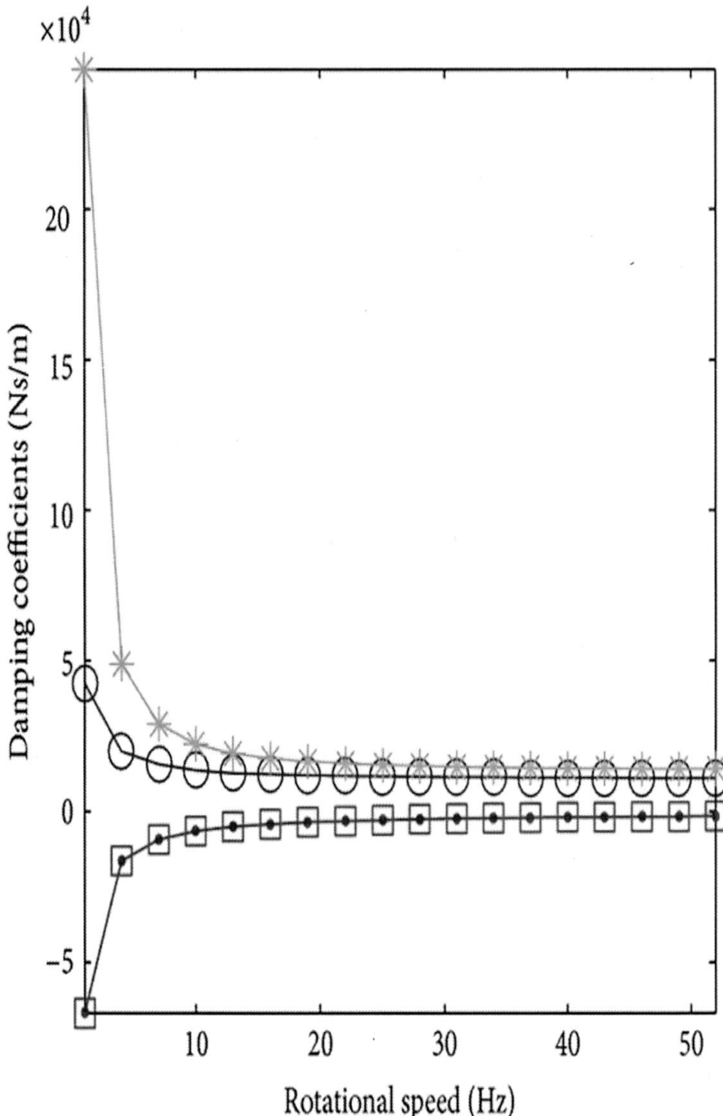

Figure 4. Equivalent coefficients of damping for the journal bearings: (a) bearing 1 in configuration A; (b) bearing 2 in configuration A; (c) bearing 1 in configuration B; (d) bearing 2 in configuration B.

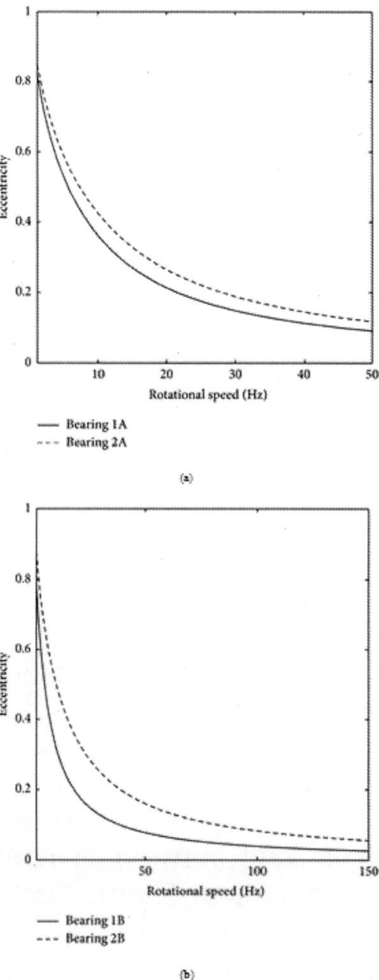

Figure 5. Journal bearings eccentricity: (a) configuration A; (b) configuration B.

It should be noticed that, although the coefficients are speed dependent, the model of the entire system (rotor and arings) is linear, what implies that the system thresholds of instability when performing run-up and run-down simulations are the same.

It is important to point out that the observance of the instability threshold in a standard rotordynamic analysis (frequency domain) is quite interesting from the practical point of view, being here the main focus of this paper.

RESULTS AND DISCUSSION

As stated before, in journal bearings supported rotors, two kinds of forward self-excited vibrations are present: oil-whirl and oil-whip [11]. As vastly discussed in the literature, oil-whip is an unstable vibration which occurs near twice the first critical speed of the rotor, occurring when oil-whirl coincides with the natural frequency of the shaft. This phenomenon was simulated and experimentally observed in a test rig used in previous works [24, 29], here called configuration A. However, when the test rig configuration was changed (configuration B) a different kind of instability was observed in the simulations and in the experimental setup. In order to better understand the different instability mechanisms, a series of simulations was proposed, which is presented in this survey.

Homogeneous Solution

From the solution of the homogeneous part of (1) it is possible to obtain the Campbell and modal damping diagrams and the natural modes of the system. Figure 6 presents the Campbell and modal damping diagrams for both configurations. In the Campbell diagram, when the rotational speed curve (1x line) crosses one of the natural frequencies of the rotor, there is a resonance region and the rotational speed, at this point, is called critical speed. Hence, a critical speed is clearly observed in configuration A (Figure 6(a)) at approximately 26.6 Hz while in configuration B (Figure 6(c)) close to 75 Hz. When simulating a shaft finite element model containing journal bearings modeled by linear coefficients of stiffness and damping, two natural frequencies (one for each bearing) can be found near half the rotational speed of the rotor (Figures 6(a) and 6(c)).

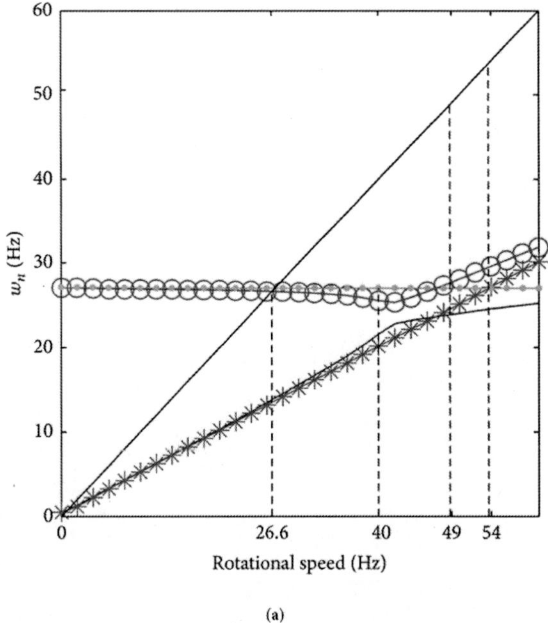
60
50
40
30
20
10
0
w_n (Hz)
0
26.6
40
49
54
Rotational speed (Hz)

(a)

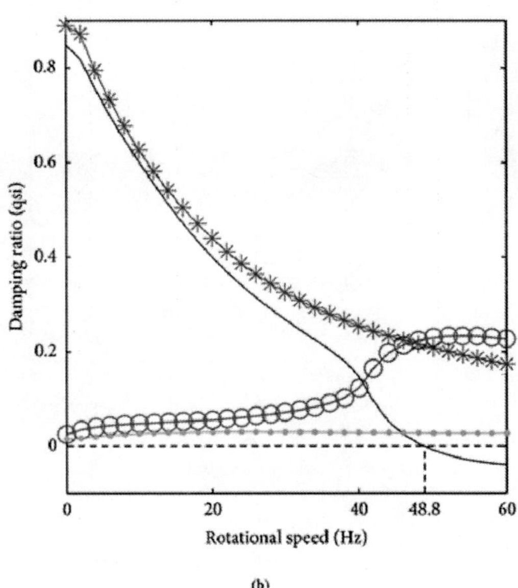
0.8
0.6
0.4
0.2
0
Damping ratio (qsi)
0
20
40
48.8
60
Rotational speed (Hz)

(b)

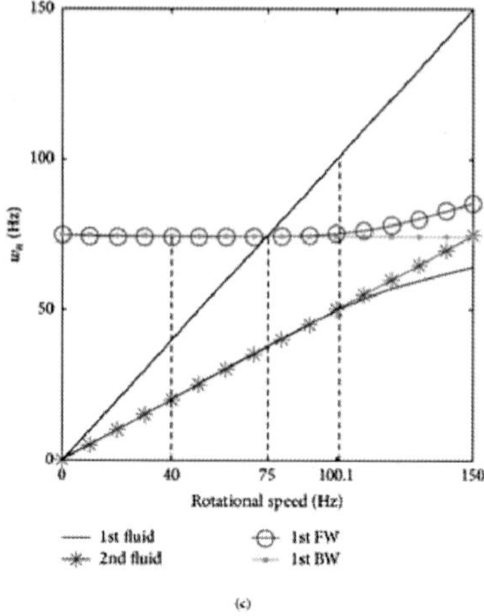

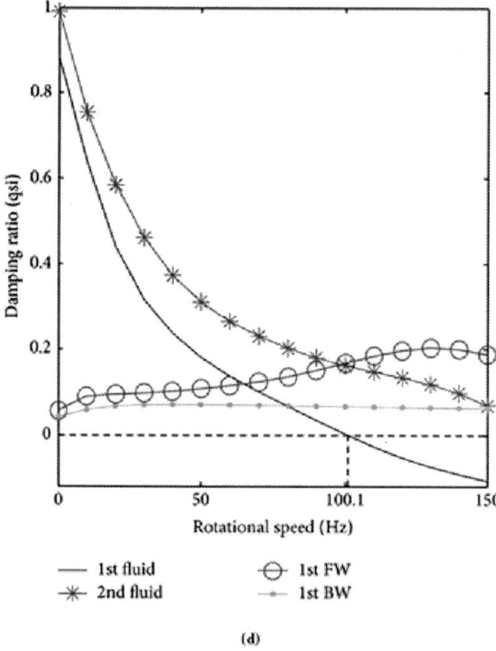

Figure 6. (a) Campbell diagram of configuration A; (b) modal damping of configuration A; (c) Campbell diagram of configuration B; (d) modal damping of configuration B.

In fact, these frequencies show up due to the effect of the journal bearings stiffness coefficients representing the fluid film. The presence of such frequencies allows the linear model to represent the fluid-induced vibrations phenomena with a very good agreement with experimental observations of rotor dynamic behaviors. Muszynska [11] simulated a flexible rotor supported by one rigid bearing and one journal bearing, modeled by the modal parameters of the first bending mode. Three eigenvalues were encountered, with two of them related to the first bending modes and one with the frequency (imaginary part) close to the oil-whirl frequency. The author noted that the real part of this third eigenvalue predicts the threshold of instability. In the words of Muszynska [11], this is an "unconventional fluid/solid interaction-related natural frequency of the rotor/bearing system," hereinafter denoted as fluid frequencies. As it was shown, in the finite element models of the rotors considered in this paper, there were two fluid frequencies, one related to each journal bearing.

Moreover, Figure 6(a) shows one of the fluid frequencies crossing the first natural frequency of the rotor when the rotational speed is of approximately 53.6 Hz, what means to excite the first mode of the rotor at a rotational speed nearly twice the critical speed of the rotor. It can also be seen, in the modal damping diagram of configuration A (Figure 6(b)), that near twice the first natural frequency of the rotor (53.6 Hz) one of the fluid frequencies damping assumes a negative value, meaning that the system became unstable after the frequency of 49 Hz. This instability threshold occurs very close to the oil-whip frequency, as it can be seen in the Campbell diagram. Once this instability threshold is achieved, its frequency does not change anymore, because the rotor enters in the oil-whip instability and it develops a high level vibration with a frequency component near its natural frequency, which does not change even if the rotor accelerates [11]. During oil-whip the bearing vibration amplitude is limited by the bearing clearance, but the rotor develops a much higher vibration level, once its natural frequency is being excited, as previously mentioned.

Analogously, Figure 6(c) shows the fluid frequency crossing the first natural frequency (75 Hz) at the rotational speed of 150 Hz, which characterizes an oil-whip frequency in this case. However, the modal damping diagram of configuration B (Figure 6(d)) shows that this configuration has an instability threshold at 101 Hz, which is much lower than the expected oil-whip instability, what suggests that a different kind of fluid-induced instability occurs. The behavior of the fluid frequency that gets unstable did not become constant as it did in configuration A, pointing to the oil-whirl as the instability cause, although it is interesting to point out that near the expected oil-whip frequency (150 Hz) the tendency of being constant is observable, suggesting that oil-whip instability would happen at that frequency if the rotor did not get unstable sooner for another reason.

The mode shapes of configuration A are shown in Figure 7, and in Figure 8, for configuration B, respectively. Figures 7(a) and 7(b) give the natural mode of the rotor at the rotational speed of 40 Hz, corresponding respectively to the fluid frequencies of 21.83 Hz and 20.85 Hz—regarding to the hydrodynamic bearings (whirl precession). Figures 7(c) and 7(d) bring the forward (25.74 Hz) and backward (27.01 Hz) natural modes of the rotor when the rotational speed is 40 Hz.

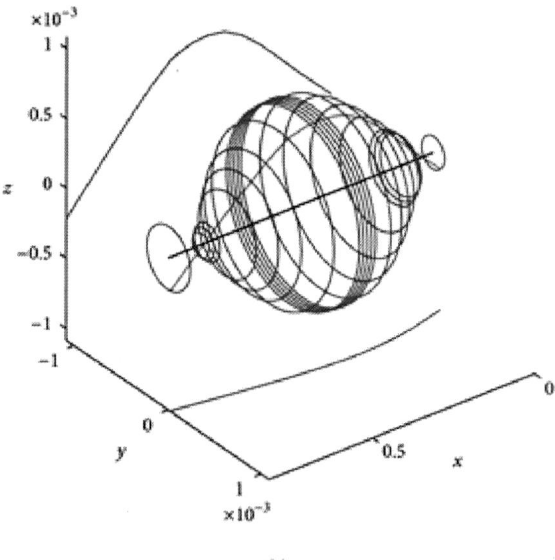

(a)

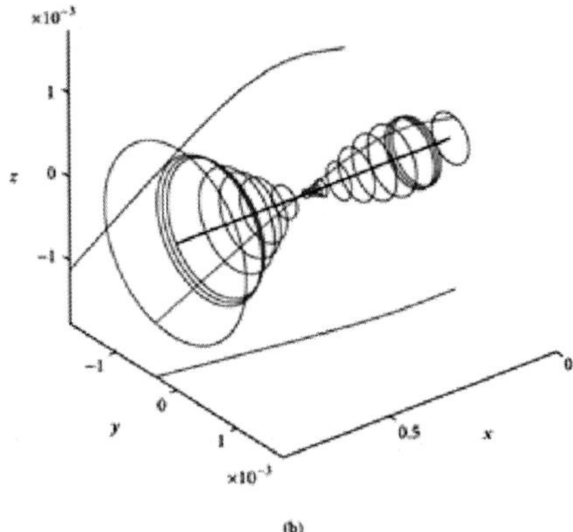

(b)

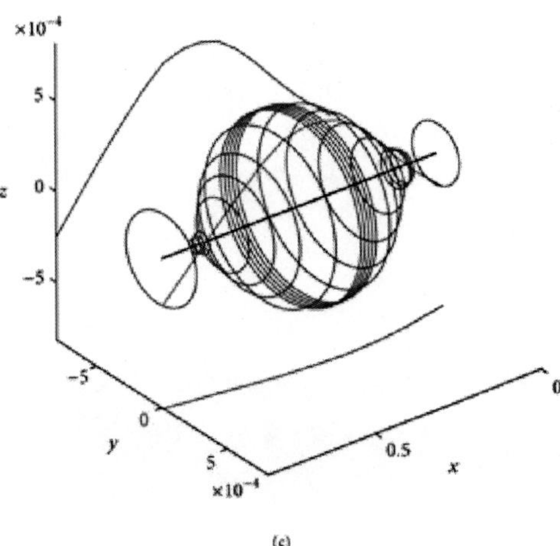

(c)

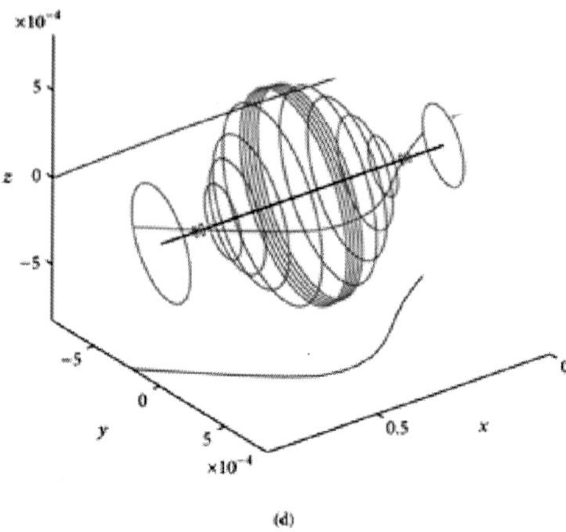

Figure 7. Natural modes of configuration A for a rotational speed of 40 Hz: (a) bearing 2 fluid frequency at 21.83 Hz; (b) bearing 1 fluid frequency at 20.85 Hz; (c) first forward flexural natural frequency at 25.74 Hz; (d) first backward flexural natural frequency at 27.01 Hz.

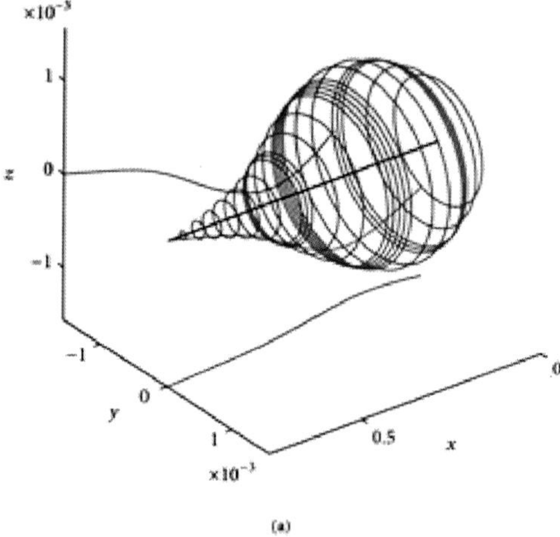

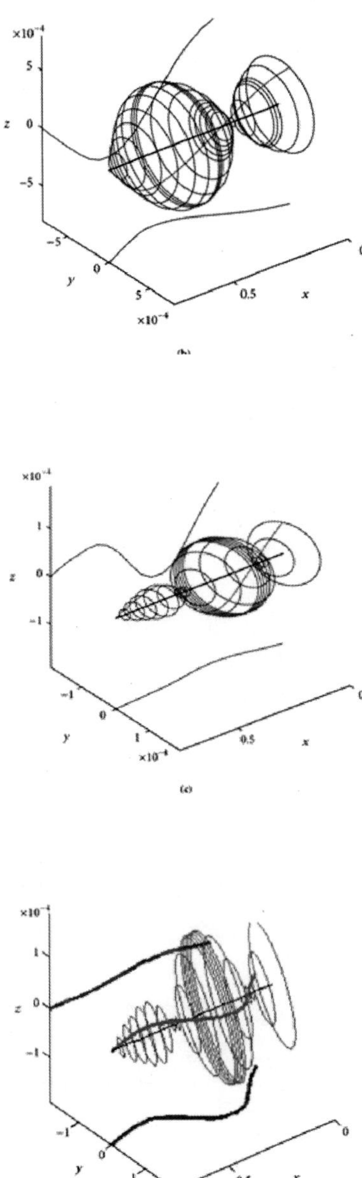

Figure 8. Natural modes of configuration B for a rotational speed of 40 Hz: (a) bearing 2 fluid frequency at 20.94 Hz; (b) bearing 1 fluid frequency at 21.85 Hz; (c) first forward flexural natural frequency at 74.89 Hz; (d) first backward flexural natural frequency at 74.882 Hz.

The mode shapes of the fluid frequencies for configuration B (Figures 8(a) and 8(b)) show that each fluid frequency corresponds to each of the journal bearings due to the fact that in each mode shape one of the bearings presents a high displacement; such effect is not so clear in configuration A (Figure 7(a)).

Up to this point, it was shown that through the homogeneous solution of the system equation, configuration A presents oil-whip instability, but configuration B presents an instability threshold lower than expected, suggesting that oil-whirl dominates the unbalance response, defining the threshold of instability.

Forced Response

In order to obtain a better understanding of the instability mechanism, run-down simulations were performed for both configurations (Figure 9). Configuration A was simulated from 50 Hz to 3 Hz, once the analysis showed the system became unstable at 49 Hz, and configuration B from 150 Hz to 3 Hz, because the modal damping diagram (Figure 6(d)) indicates that instability starts at 101 Hz, but the Campbell has evidence that oil-whip would happen nearly at 150 Hz. In both cases, a deceleration of 1 Hz/s was used.

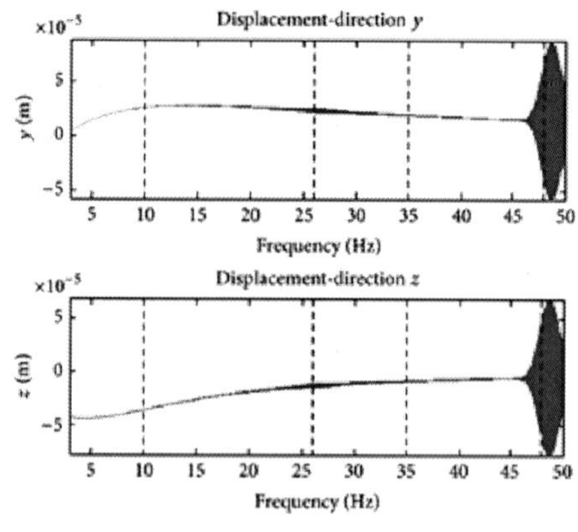

(a)

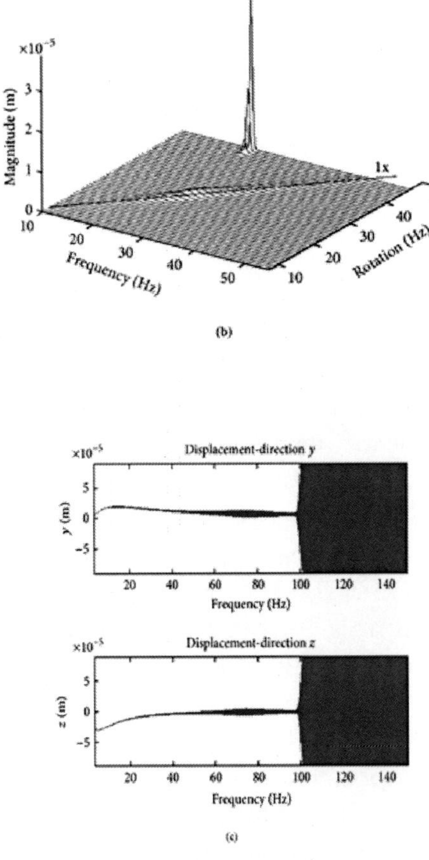

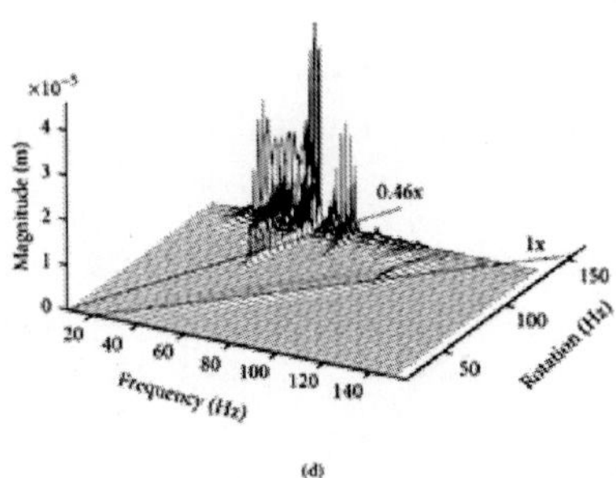

Figure 9. Run-down simulations: (a) time response of the first bearing in configuration A (see Figure 1); (b) waterfall diagram of the first bearing in configuration A; (c) time response of the first bearing in configuration B; (d) waterfall diagram of the first bearing in configuration B.

Remark

The reason why run-down simulations were used instead of run-ups is related to the instability sensitivity of the linear model used for the journal bearings. It was noticed in run-up simulations that, even though the rotor was unstable after the instability threshold, the vibration amplitude rises in a very slow shape. Consequently, a run-down simulation may be set to start in a rotor unstable condition, with large vibration amplitudes, overcoming the run-up simulation limitation. Hence, run-down simulations are suitable to the study of fluid-induced instability simulations in cases where, like in this work, the entire system model is linear (recalling that the journal bearings were modeled using a linear coefficients approach), and the same instability thresholds are obtained either for run-up or run-down simulations. It should also be noticed that the journal bearings nodes vibration is limited by the bearing radial clearance. The interaction between the journal and the housing wall is not in the scope of this paper. Thereby, a simple approach was used: when the vibration amplitude gets higher than the radial clearance, the angular position of the shaft inside the bearing is kept constant and its amplitude is reduced to the clearance value. The disadvantage of this approach is that harmonic components showed up in the waterfall diagrams. Hence, their influence should be disregarded in the analysis.

Figures 9(a) and 9(c) show the time response for the run-down in the first journal bearing of each configuration, where it is clear that the systems were unstable once the stability thresholds were crossed. The self-centering effect of the shaft due to unbalance (reduction of the orbit amplitude after the critical speed) and the self-centering effect of the bearing can also be noticed.

Figures 9(b) and 9(d) contain the waterfall diagrams correspondent to Figures 9(a) and 9(c), respectively. Oil-whirl did not show up in configuration A (Figure 9(b)). The system presents only the unbalance response until oil-whip takes place, when the modal damping of the fluid frequency crosses the zero. In real rotors, it can be seen that the orbit of the shaft inside the bearings presents a deployment due to a sub harmonic component at half the rotational speed (i.e., oil-whirl); and when the rotor rotational speed has a small increment and crosses the instability threshold, the oil-whip becomes dominant [11]. Figure10 presents the orbits and the spectra of the first bearing in several constant rotational speeds (dashed lines in Figure 9(a)), before the critical speed (10 Hz), at the critical speed (26.6 Hz), between the critical speed and the instability threshold (35 Hz), and at 48.8 Hz (close to the instability), and no deployment can be seen, which demonstrate a limitation of the linear model once this phenomenon was observed in the nonlinear model used in the work of Castro et al. [24], although the instability threshold presented good agreement. The large orbit seen in Figure 9(a) at 48 Hz is a transient effect of the system going out of the instability region (near 48.8 Hz), once this result was obtained from a run-down simulation as previously mentioned.

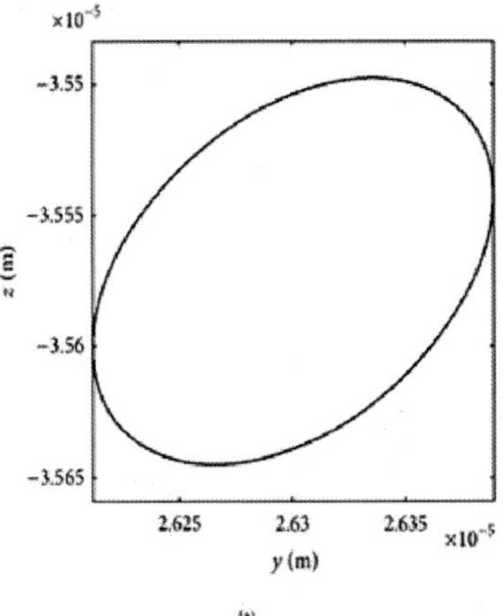

(a)

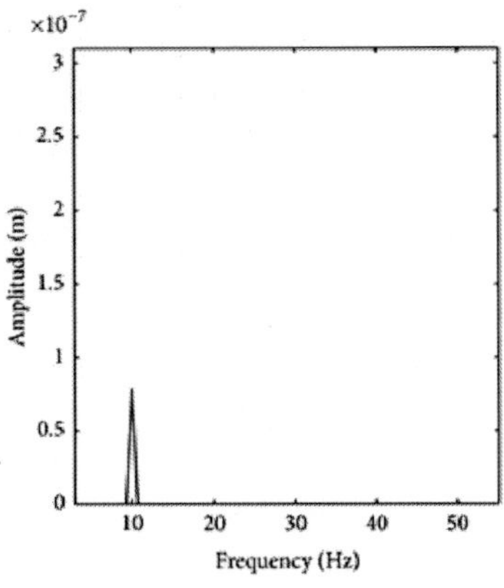

(b)

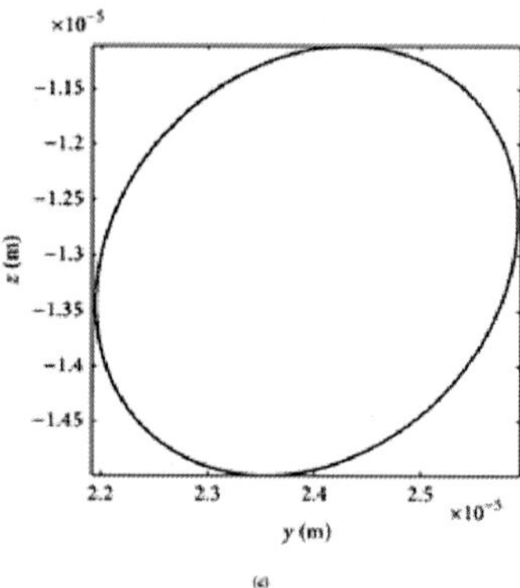
×10⁻⁵
−1.15
−1.2
−1.25
−1.3
−1.35
−1.4
−1.45
z (m)
2.2
2.3
2.4
2.5
×10⁻⁵
y (m)
(c)

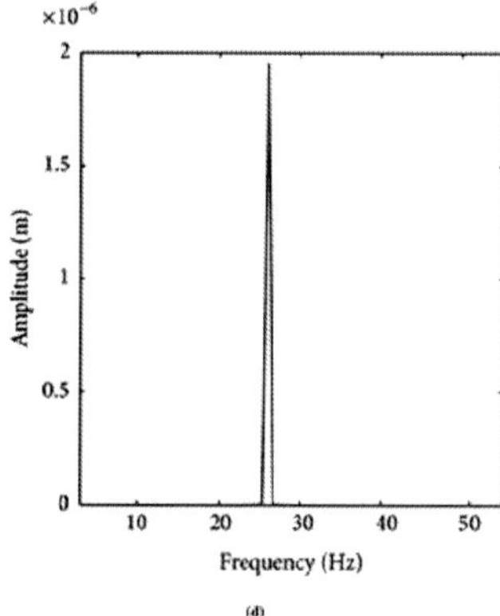
×10⁻⁶
2
1.5
1
0.5
0
Amplitude (m)
10
20
30
40
50
Frequency (Hz)
(d)

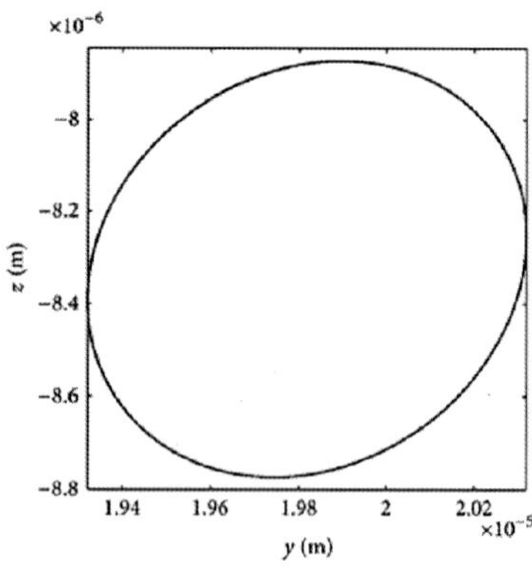
×10⁻⁶
−8
−8.2
−8.4
−8.6
−8.8
z (m)
1.94
1.96
1.98
2
2.02
×10⁻⁵
y (m)
(e)

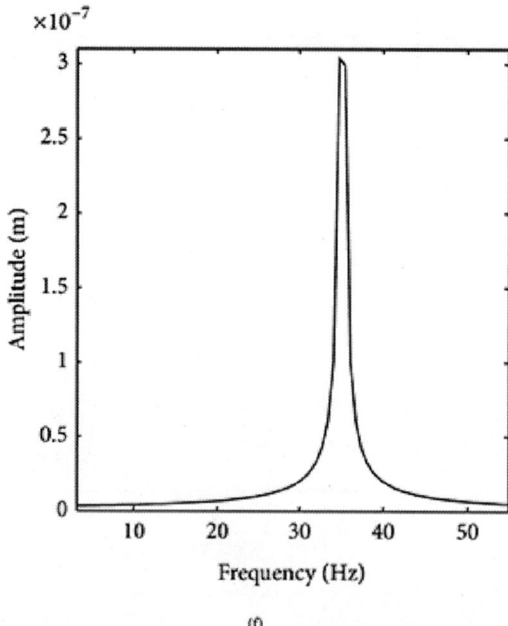
×10^-7
3
2.5
2
1.5
1
0.5
0
Amplitude (m)
10
20
30
40
50
Frequency (Hz)

(f)

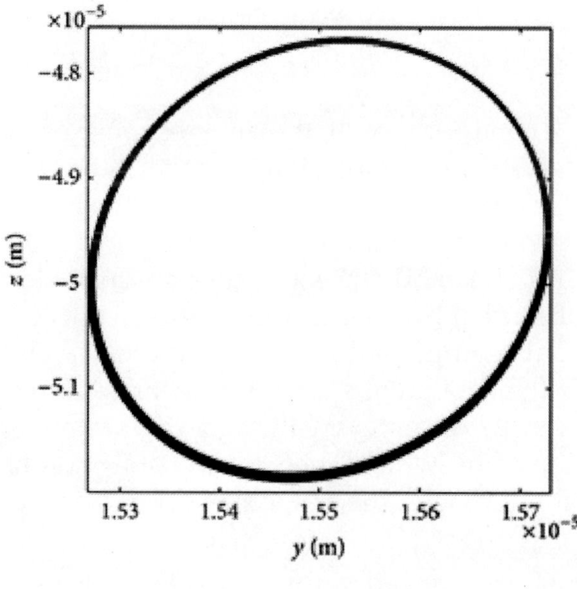
×10^-5
−4.8
−4.9
−5
−5.1
z (m)
1.53
1.54
1.55
1.56
1.57
×10^-5
y (m)

(g)

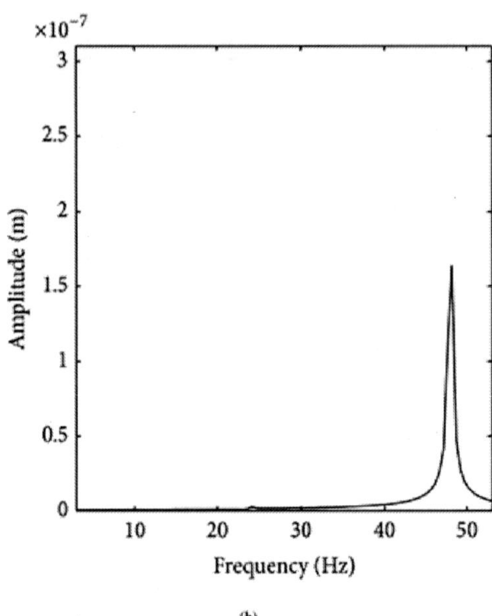

Figure 10. Simulated results of the first bearing in configuration A for constant rotational speeds (see Figure 8(a)): (a) orbit at 10 Hz; (b) frequency spectrum at 10 Hz; (c) orbit at 26 Hz; (d) frequency spectrum at 26 Hz; (e) orbit at 35 Hz; (f) frequency spectrum at 35 Hz; (g) orbit at 48 Hz; (h) frequency spectrum at 48 Hz.

In the case of configuration B, the time response of the run-down (Figure 9(c)) also shows the instability of the rotor, when the orbit amplitude is limited by the bearing clearance. However, in the waterfall diagram it can be seen that near the instability threshold (101 Hz) oil-whirl rises and starts to cause an unstable vibration. It is possible to observe it because as the rotational speed changes, the unstable harmonic component of the vibration also changes, being close to 46% of the rotational speed until nearby 150 Hz, when oil-whip is expected to occur even though the system was already unstable because of oil-whirl. Thereby, in configuration B, instead of oil-whip, oil-whirl was the responsible for driving the system unstable.

Constant rotational speed simulations were also performed for two cases in configuration B, at 99 Hz and close to the instability limit at 101 Hz; the orbits of the first bearing and the frequency spectrum are presented in Figure 11. The experimental results, which have motivated this survey, are also shown in Figure 11 overlaid to the simulated results to highlight the differences. All measurements were made using 3300 RAM inductive sensors from Bently Nevada.

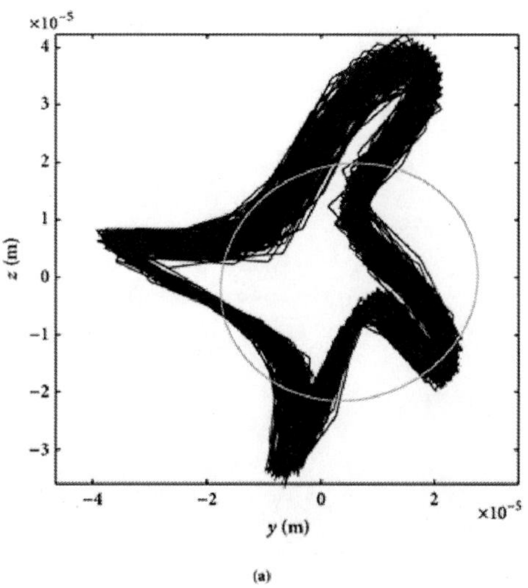

(a)

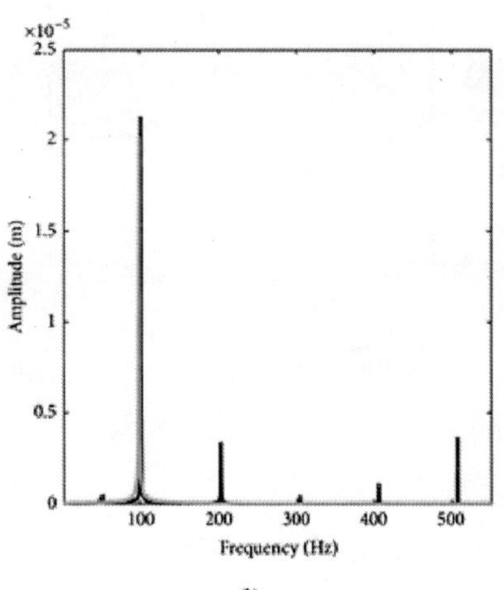

(b)

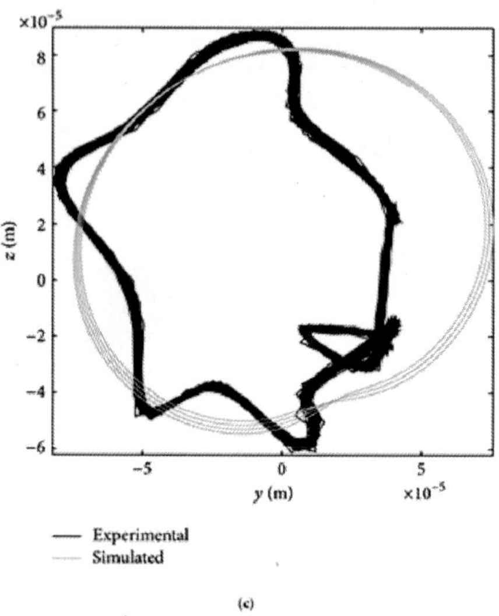

(c)

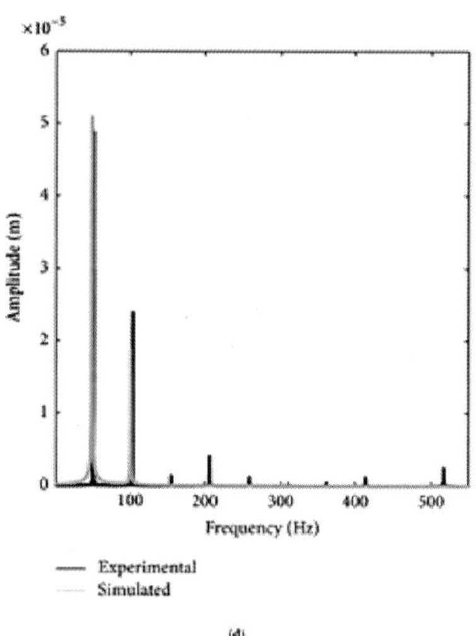

(d)

Figure 11. Experimental and simulated results for two conditions with fixed rotational speeds in configuration B: (a) orbits of the first condition (experimental 99 Hz, simulated at 101 Hz); (b) frequency spectrum of the first condition (experimental at 99 Hz, simulated at 101 Hz); (c) orbits of the second condition (experimental at 101.4 Hz, simulated at 103.4 Hz); (d) frequency spectrum of the second condition (experimental at 101.4 Hz, simulated at 103.4 Hz).

At the rotational speed of 99 Hz (gray curves in Figures 11(a) and 11(b)) only the unbalance effect appears, but as the rotational speed was increased to 101 Hz, the orbit deployment appears (Figure 11(c)in gray), indicating the presence of oil-whirl as it can be seen in the frequency spectrum (Figure 11(d) in gray).

The experimental threshold of instability is slightly shifted to 103.4 Hz. Therefore, the first difference between experiment and simulation is in the constant rotational speeds measured, 100.4 Hz and 103.4 Hz, which shows that the model is a little bit more conservative. The orbit shapes (black curves in Figures 11(a) and 11(c)) are considerably different from the simulated results, and the reason is the presence of many harmonics found in the experimental frequency spectra (Figures 11(b) and 11(d) in black). Such harmonics occur due to the ball bearing at the shaft end in configuration B; as the rotor spins and its spheres pass through the lower part of the bearing, where the load is applied, harmonics of the rotational speed are excited and participate in the vibration response. This effect is not present in the simulations once the ball bearing is merely modeled by constant stiffness and damping coefficients (Figure 11(b)). However, the rotational speeds associated with the threshold of instability are very close and the oil-whirl components in the spectra present about the same amplitudes obtained by the simulation response (Figure 11(d)).

Therefore, comparing the two different cases, it can be seen that in one case the instability was caused by oil-whirl and, in the other case, by oil-whip. It should be reminded at this point that oil-whirl could exist in a stable condition until it turns to oil-whip, as presented in Castro et al. [24]; such behavior would be expected by a lighter rotor than the ones used in the present survey. Moreover, it was shown here that the instability threshold could be safely estimated when using the linear coefficients approach for journal bearings modeling.

Eccentricity Analysis

The denominations heavy rotor or light rotor used when talking about horizontal rotors supported on journal bearings are related to the eccentricity of the shaft during operation. The bearing eccentricity is a function of the load applied on each bearing and the bearing load capacity at each rotational speed. In this sense, the terminology heavy rotor and light rotor refers to the relation between bearing load and load bearing capacity and how it affects the bearing eccentricity. The bearing load is constant and dependent on the geometry of the rotating system; yet, as bearing load capacity is a function of the rotational speed, so is the bearing eccentricity.

When operating at low rotational speeds, a journal bearing pumps a little quantity of oil underneath its journal, resulting in a thin stiff fluid-film and in a high eccentricity; thus, oil-whirl is not present and the rotor is stable.

As the rotational speed increases, the decreasing in the eccentricity and the rising of the attitude angle (tending to 90 degrees) results in the self-centering effect. The thick fluid film causes disequilibrium of the stabilizing forces inside the bearings; the same condition of low eccentricity occurs in vertical or lightly loaded horizontal rotors as they operate nearly centered.

Figure 12 presents the forces acting on the shaft inside the bearing as terms of the stiffness and damping coefficients. In order to maintain stable operation, the dissipative force (and) must suppress the destabilizing effect of the cross-coupled stiffness (and). However, as the rotational speed increases, the cross-coupled stiffness coefficients increase drastically, while the direct stiffness and the damping coefficients present only small changes (Figures 3 and 4), which causes a reduction of stability and oil-whirl may appear.

Consequently, the different behaviors of the instability mechanism in both configurations A and B are due to the relation between bearing load and bearing load capacity. If the rotor has one bearing working in a light rotor condition, such that it became unstable before oil-whirl reaches the natural frequency of the rotor (and turn to oil-whip), then oil-whirl will be the cause of the instability. If the bearings work in a heavy rotor condition (higher eccentricity), then oil-whip will cause the instability at twice the rotor natural frequency. Oil-whirl instability threshold here is considered to be the point where the subsynchronous harmonic is limited only by the bearing clearance; and this effect is observed in the linear models as a negative value associated with the fluid-frequency in the modal diagrams.

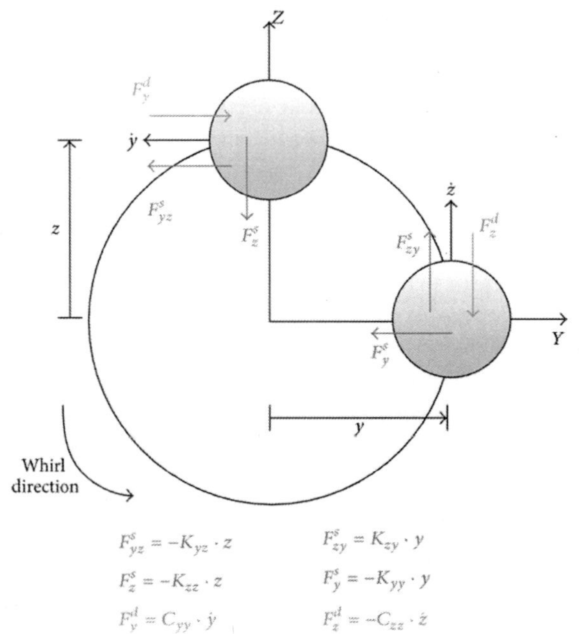

Figure 12. Representation of the forces acting on the shaft inside the journal bearing.

If the rotor keeps on accelerating, it will achieve the instability threshold and oil-whirl may become unstable (as in configuration B) or it may give place to oil-whip (as in configuration A).

Since the same bearings were used in both cases, the load in each case tells which bearing is more unstable. The loads for configuration A are 17.7 N for the first bearing and 23.1 N for the second, and configuration B has 12.5 N on the first bearing and 26.7 N on the second. This means that the rotor from configuration B has a bearing with a much lower eccentricity (less stable) as it can be seen in Figure 5, what explains why the instability mechanism has a different behavior from the previous configuration. In configuration B, oil-whirl became unstable much sooner than the expected oil-whip. Only for comparison, the load in the ball bearing in configuration B is 1.46 N; therefore, the sum of the loads of the bearings in each configuration allows observing that both configurations have almost the same weight (40.8 N for configuration A and 40.66 N for configuration B). This fact agrees with the early explanation between how the rotor weight is not the only parameter related to the light/ heavy rotor denominations, but it strongly depends on the load applied to each bearing and the bearing load capacity.

CONCLUSIONS

Motivated by experimental results, a survey about the fluid-induced instability mechanism was presented considering two different rotor configurations. The hydrodynamic bearings were modeled as linear coefficients of stiffness and damping, evaluated through the solution of the Reynolds equation using a finite-volume method. The first configuration instability threshold is the expected oil-whip frequency; however, in the second configuration, the instability threshold occurs much sooner than expected, being caused by an unstable oil-whirl vibration. The analysis and comparison of both cases were made through Campbell and modal damping diagrams and, afterwards, run-down simulations (waterfall diagram) validation. In both cases, the instability threshold and the instability cause (oil-whirl or oil-whip) could be successfully identified through the models, presenting good agreement with experimental results even though no model adjustment techniques were used. Hence, the linear coefficients approach could be used in the design stage as a first indicative of the machine instability limit. Finally, a discussion about the instability mechanism and how it is affected by the bearing load and the bearing load capacity was presented.

APPENDIX

The matrixes used to build the finite element model are given bywhere the area moment of inertia is given by

NOTATION

Roman Letters

d_e: Shaft element diameter
h: Fluid film thickness
L_e: Shaft element length
p: Pressure
$\mathbf{q}$: Generalized coordinate vector
t: Time
x, r: Axial and circumferential coordinates of the bearing
y, z: Coordinates of the plane perpendicular to the shaft
A: Cross-section area of the shaft element
C: Damping coefficient
$\mathbf{C}$: Damping matrix
E: Young modulus
$\mathbf{F}$: Force vector
$\mathbf{G}$: Gyroscopic matrix
I_{yy}, I_{zz}: Area moments of inertia
K: Stiffness coefficient
$\mathbf{K}$: Stiffness matrix
$\mathbf{M}$: Mass matrix
$\mathbf{M_T}$: Translational mass matrix
$\mathbf{M_R}$: Rotational mass matrix.

Geek Letters

Λ: Infinitesimal variation
ρ: Shaft material density
μ: Fluid viscosity
Ω: Rotational speed.

Superscripts

$\cdot$: Differentiation with respect to time
d: Force due to a damping coefficient
s: Force due to a stiffness coefficient.

Subscript

0: Equilibrium position.

CONFLICT OF INTERESTS

The authors declare that there is no conflict of interests regarding the publication of this paper.

ACKNOWLEDGMENTS

The authors would like to thank the funding support by CNPq (ConselhoNacional de DesenvolvimentoCientífico e Tecnológico) (Grantno.131658/2010-7);CAPES(Coordenação de Aperfeiçoamento de Pessoal de Nível Superior) (PROBRAL 341/10); and FAPESP (Fundação de Amparo à Pesquisa do Estado de São Paulo) (Grant no. 2007/54647-4).

REFERENCES

1. N. P. Petrov, "Friction in machines and the effect of lubricant," InzhenernyjZhurnal, Sankt-Peterburg, vol. 1, pp. 71–140, 1883.
2. N. P. Petrov, "Friction in machines and the effect of lubricant," InzhenernyjZhurnal, Sankt-Peterburg, vol. 2, pp. 228–279, 1883.
3. N. P. Petrov, "Friction in machines and the effect of lubricant," InzhenernyjZhurnal, Sankt-Peterburg, vol. 3, pp. 377–436, 1883.
4. N. P. Petrov, "Friction in machines and the effect of lubricant," InzhenernyjZhurnal, Sankt-Peterburg, vol. 4, pp. 535–564, 1883.
5. B. Tower, "First report on friction experiments (friction of lubricated bearings)," pp. 632–659, 1883, (Adjourned Discussion, pp. 29–35, 1884).
6. B. Tower, "Second report on friction experiments (experiments on the oil pressure bearing),"Proceedings of the Institution of Mechanical Engineers, pp. 58–70, 1885.
7. O. Reynolds, "On the theory of lubrication and its application to Mr. Beauchamp tower's experiments, including an experimental determination of the viscosity of olive oil," Philosophical Transactions of the Royal Society A, vol. 177, no. 1, pp. 157–234, 1886.
8. G. Oliver, "An Introduction to Oil Whirl and Oil Whip," Turbo Components and Engineering Newsletter: Bearing Journal, vol. 3, no. 2, pp. 1–2, 2001.
9. B. L. Newkirk, "Shaft Whipping," General Electric Review, vol. 27, article 169, 1924.

10. B. L. Newkirk and H. D. Taylor, "Shaft whipping due to oil action in journal bearings," General Electric Review, vol. 28, pp. 559–568, 1925.
11. A. Muszynska, "Whirl and whip—rotor/bearing stability problems," Journal of Sound and Vibration, vol. 110, no. 3, pp. 443–462, 1986. View at Scopus
12. A. Muszynska, "Stability of whirl and whip in rotor/bearing systems," Journal of Sound and Vibration, vol. 127, no. 1, pp. 49–64, 1988. View at Scopus
13. A. Stodola, "KritischeWellenstörunginfolge der Nachgiebigkeit des Oelpolstersim Lager,"SchweizerischeBauzeitung, vol. 85-86, pp. 265–266, 1925 (German).
14. C. Hummel, Kritischedrehzahlenalsfolge der nachgiebigkeit des schmiermittelsim lager [Thesis], EidgenössischenTechnischenHochschule, Zurich, Switzerland, 1926 (German).
15. J. W. Lund, "Spring and damping coefficients for the tilting pad journal bearing," ASLE Transactions, vol. 7, no. 4, pp. 342–352, 1964. View at Publisher · View at Google Scholar
16. T. H. Machado and K. L. Cavalca, "Evaluation of dynamic coefficients of fluid journal bearings with different geometries," in Proceedings of the 20th Brazilian congress of mechanical engineering (COBEM '09), ABCM, Gramado, Brazil, November 2009.
17. T. H. Machado, Evaluation of hydrodynamic bearings with geometric discontinuities [Dissertation], University of Campinas, Campinas, Brazil, 2011 (Portuguese).
18. G. Capone, "Orbital motions of rigid symmetric rotor supported on journal bearings," La Meccanicaltaliana, vol. 199, pp. 37–46, 1986 (Italian).
19. G. Capone, "Analytical description of fluid-dynamic force field in cylindrical journal bearing,"L'EnergiaElettrica, vol. 3, pp. 105–110, 1991 (Italian).
20. J. M. Vance, Rotordynamics of Turbomachinery, John Wiley & Sons, New York, NY, USA, 1st edition, 1988.
21. E. Krämer, Dynamics of Rotors and Foundations, Springer, New York, NY, USA, 1st edition, 1993.
22. D. Childs, TurbomachineryRotordynamics: Phenomena, Modeling, and Analysis, Wiley-Interscience, New York, NY, USA, 1st edition, 1993.
23. E. N. Lima, Non-linear model for hydrodynamic sustaining forces on vertical rotors journal bearings [Dissertation], University of Campinas,

Campinas, Brazil, 1996 (Portuguese).

24. H. F. de Castro, K. L. Cavalca, and R. Nordmann, "Whirl and whip instabilities in rotor-bearing system considering a nonlinear force model," Journal of Sound and Vibration, vol. 317, no. 1-2, pp. 273–293, 2008. View at Publisher · View at Google Scholar · View at Scopus
25. H. D. Nelson and J. M. McVaugh, "The dynamics of rotor-bearing systems using finite elements,"Journal of Engineering for Industry, vol. 98, no. 2, pp. 593–600, 1976. View at Publisher · View at Google Scholar
26. A. A. G. Siqueira, R. Nicoletti, N. Norrick et al., "Linear parameter varying control design for rotating systems supported by journal bearings," Journal of Sound and Vibration, vol. 331, no. 10, pp. 2220–2232, 2012. View at Publisher · View at Google Scholar · View at Scopus
27. B. Riemann, E. A. Perini, K. L. Cavalca, H. F. Castro, and S. Rinderknecht, "Oil whip instability control using μ-synthesis technique on a magnetic actuator," Journal of Sound and Vibration, vol. 332, no. 4, pp. 654–673, 2013. View at Publisher · View at Google Scholar
28. J. Tuma, J. Simek, J. Skuta, and J. Los, "Active vibrations control of journal bearings with the use of piezoactuators," Mechanical Systems and Signal Processing, vol. 36, no. 2, pp. 618–629, 2013.View at Publisher · View at Google Scholar
29. R. U. Mendes, L. O. S. Ferreira, and K. L. Cavalca, "Analysis of a complete model of rotating machinery excited by magnetic actuator system," Proceedings of the Institution of Mechanical Engineers C, vol. 227, no. 1, pp. 48–64, 2013. View at Publisher · View at Google Scholar
30. P. M. Santana, K. L. Cavalca, E. P. Okabe, and T. H. Machado, "Complex response of a rotor-bearing-foundation system," in Proceedings of the 8th IFToMM International Conference on Rotordynamics, vol. 1, pp. 1–8, KAIST University, Seoul, Republic of Korea, September 2010.
31. T. H. Machado and K. L. Cavalca, "Dynamic analysis of cylindrical hydrodynamic bearings with geometric discontinuities," in Proceedings of the 10th International Conference on Vibration Problems (ICOVP '11), J. Náprstek, J. Horáček, M. Okrouhlík, B.

Marvalová, F. Verhulst, and J. T. Sawicki, Eds., vol. 139 of Proceedings in Physics, pp. 537–542, Springer, Prague, Czech Republic, September 2011.

32. J. W. Lund, "Review of the concept of dynamic coefficients for fluid film journal bearings,"Journal of Tribology, vol. 109, no. 1, pp. 37–41, 1987. View at Scopus

Chapter 2

EXPERIMENTAL STUDY OF DARRIEUS-SAVONIUS WATER TURBINE WITH DEFLECTOR: EFFECT OF DEFLECTOR ON THE PERFORMANCE

Kaprawi Sahim, Kadafilhtisan, Dyos Santoso, and Riman Sipahutar

Mechanical Engineering, Sriwijaya University, Jalan Raya Palembang-Prabumulih Km 32, Indralaya 30662, Indonesia

ABSTRACT

The reverse force on the returning blade of a water turbine can be reduced by setting a deflector on the returning blade side of a rotor. The deflector configuration can also concentrate the flow which passes through the rotor so that the torque and the power of turbine can be considerably increased. The placing of Savonius in Darrieus rotor is carried out by setting the Savonius bucket in Darrieus rotor at the same axis. The combination of these rotors is also called a Darrieus-Savonius turbine. This rotor can improve torque of turbine. Experiments are conducted in an irrigation canal to find the performance characteristics of presence of deflector and Savonius rotor in Darrieus-Savonius turbine. Results conclude that the single deflector plate placed on returning blade side increases the torque and power coefficient. The presence of Savonius rotor increases the torque at a lower speed, but the power coefficient decreases. The torque and power coefficient characteristics depend on the aspect ratio of Savonius rotor

INTRODUCTION

Exploitation of renewable energies has been intensified to meet the electricity needs when there is a potential source of hydropower. This kind of energies will decrease the environmental problems of pollution. River's flow, canal flow, irrigation flow, and tidal current still have many difficulties to be extracted optimally. This is due to the low efficiency of the energy technology in which the water turbines used for stream flow have to be developed. One of the turbines commonly used is a helical type which gives the efficiency of 35% [1]. However, if a large capacity of water flows, generating power depends on taking lots of flow rate into the turbine.

Different designs of water current turbines are available for the extraction of energy from the river water or canals. Based on the alignment of the rotor axis with respect to water flow, two generic classes exist. They are horizontal and vertical axis turbine. The commonly used vertical axis turbines are Darrieus, Gorlov, and Savonius turbine. Helical blade and Darrieus straight blade water turbine are commonly suitable for extraction of kinetic energy of flowing water. Helical blades have a smaller rate of pulsation [2] and more favorable starting characteristics than straight blades [3], but the blades are very difficult to be constructed than the straight blade.

The turbine system should be a simple structure and with a good reliability to fulfil the power generation so that one can construct by itself like for small-scale applications of local production of electricity [4]. The tip speed ratio (TSR) of Darrieus turbine is high and that is why it rotates much faster. The Darrieus type has a cost-advantage due to the simple structure. Installing the Darrieus turbine in a narrow canal with small clearance will increase the efficiency [5]. Savonius turbine is another type of simple one. This turbine has the advantage of self-starting torque, but the efficiency and the operation speed are lower. Savonius design uses a rotor that is formed by cutting the flattener cylinder into two halves along the central plane and then moving the two semicylindrical surfaces sideways along the cutting plane so that the cross-section resembled the letter "S" [6].

The efficiency of Savonius turbine can be increased by making configuration of single stage, two-stage, and three-stage rotor. These configurations have been tested in a water channel as it was conducted by Khan et al. [7]. The use of two deflector plates can increase significantly the power coefficient of modified Savonius rotor as studied experimentally by Golecha et al. [8],where one deflector is placed on the returning blade side and the other is placed on the outside of the advancing blade side. To improve the starting torque of the Darrieus turbine used for power generation, a Darrieus-Savonius turbine, composed of a Darrieus turbine and Savonius rotor, has been experimentally tested in an open channel. The Darrieus-Savonius turbine has a higher torque at a lower speed than the solo Darrieus turbine, but the efficiency is lower than the Darrieus rotor regardless of Savonius bucket orientation (attachment angle). In Darrieus-Savonius turbine, the best attachment angle of Savonius rotor is placed perpendicularly to the Darrieus rotor. The operational tip speed ratio of solo Savonius turbine is less than unity and the solo Darrieus is greater than unity. The combined turbine of Darrieus-Savonius has the operational tip speed ratio greater than unity [9]. In Darrieus-Savonius rotor, the bucket of semielliptic cylinders is used to improve the torque at a low speed; however, the rotation and efficiency are still lower than solo Darrieus turbine [10]. The usage of guide vane increases the performance of wind turbine of Darrieus type as it was tested by Takao et al. [11].

From the literature survey, the effects of deflector and the bucket size on the performance have not been clarified so far. It is the reason, in the present paper, we propose to observe experimentally the impact of the Darrieus-Savonius turbine using a deflector plate and variable bucket dimensions on the performance which is represented by limits of torque and power coefficient. The combined Darrieus-Savonius turbine consists of Darrieus and Savonius turbine in which the Savonius rotor is placed on the middle of Darrieus rotor on the same shaft. Savonius water turbine and cross-flow water turbine can be improved by setting deflector around the rotors because the deflector in upstream of the rotor can increase torque and power coefficient.

EXPERIMENTAL SETUP AND PROCEDURE

In Darrieus-Savonius turbine, the Darrieus turbine has been used as a main device and the Savonius turbine as a startup device. They are permanently attached to the same axis. The Savonius turbine is composed of two semicircular buckets. Figure 1 shows the setting of a deflector plate to the combined rotor. The deflector plate placed upstream to the fluid flow on the returning blade side acts as an obstacle to the flow coming towards the returning blade. This reduces the negative or reverse torque on the returning blade. Experiments are conducted for two positions of the deflector plate on the returning blade side of the combined rotor. In the first position, deflector is set at (without deflector) and in second position, the deflector is set at . The deflector plate has length of 100 cm. Setting the second angle is to obtain the height of the deflector . The gap between rotor blade and end plate of the deflector was about 40 mm. Figure 2 gives the dimension of the blade and bucket. The tested Darrieus rotor with the diameter of mm has two straight blades with a chord length of mm and length of mm. We use blade profile of NACA 0015 for the reason of small drag when compared to blades of higher thickness. The Darrieus-Savonius rotor has also two Savonius buckets with the diameter of a semicircular cylinder and length .

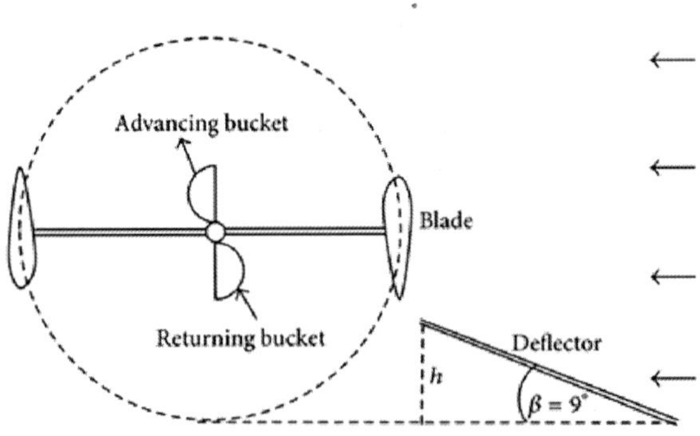

Figure 1: Darrieus-Savonius rotor.

Figure 1: Darrieus-Savonius rotor.

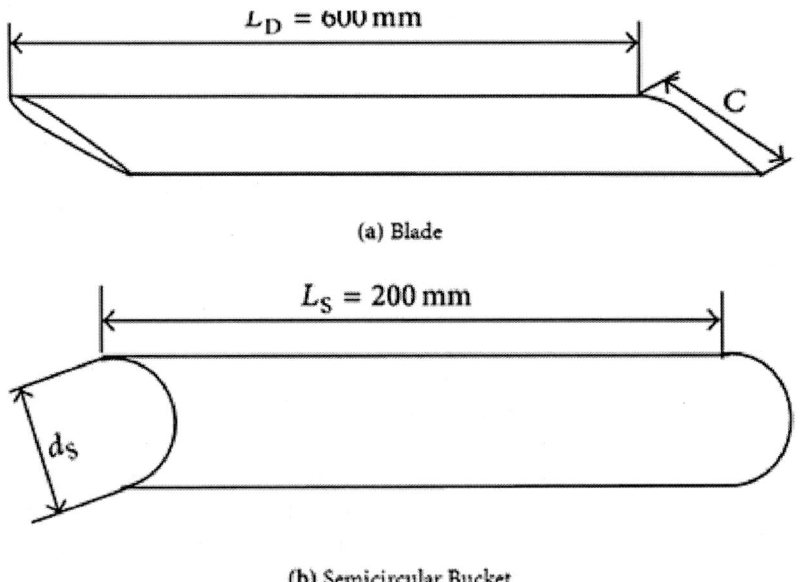

Figure 2: Geometrical blade and bucket.

One of the purposes of the study is to reveal the influence of a swept area of Savonius bucket. For that reason, we chose only two buckets of diameter mm and 63 mm in Darrieus-Savonius turbine to be tested. The aspect ratio of Savonius rotor is defined by a ratio of the length of bucket and the total width or diameter of two buckets. In this case, the total diameter of buckets is twice the diameter of each bucket. The above dimensions of the bucket give the aspect ratio of and 1.58. It is noted that higher aspect ratio represents a smaller swept area (smaller). The bucket is made of polyvinyl chloride (PVC) material and the blade of Darrieus rotor was made of wood with smooth surface. In this study, solidity of Darrieus rotor is kept constant at 0.2 and it is calculated by the following expression:

The turbine axis of diameter 14 mm is held by two ball bearings. The torque of the rotor was measured by a rope brake dynamometer with nylon rope of diameter 2 mm. The water turbine was first revolved without any loads and at this condition, the revolution was maximum. The loads were gradually added to the revolution by adding a certain mass on weighing pan; then the rotation and the load on spring scale were measured until the water turbine was completely stopped (overload). The mass on weighing pan was measured by digital scale with accuracy 500 gram/0.01 gram, while the rotational speed of the rotor was measured using a laser tachometer. The same procedures of measurement were used for the other configurations of the combined turbine. The current meter flowatch FL-03 with accuracy 2% was used to measure the free stream velocity. The tubular spring scale of range 0–50 N was used to measure the force (see Figure 3).

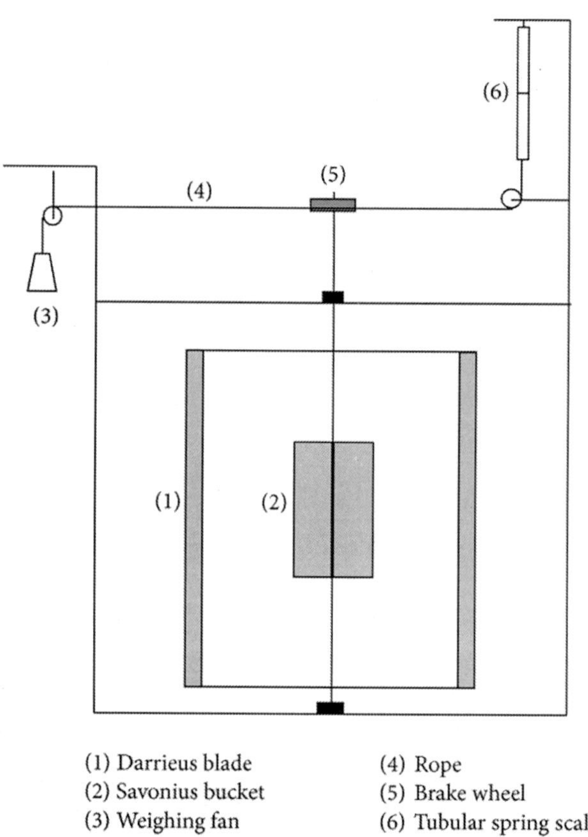

Figure 3: Schematic diagram of the test setup for the Darrieus-Savonius turbine.

The rotational speed of the rotor is represented by dimensionless parameter tip speed ratio (TSR or). TSR is calculated by the following relation:The effective torque, , is calculated byTorque coefficient is obtained from the following relation:where is available torque; is effective torque.

The performance of the turbine is given by the power coefficient and torque coefficient . These coefficients represent the produced energy of the turbine as part of the total water energy passing through the swept area of the turbine rotor. This area equals the frontal area of the turbine given by the height times the diameter. As the blades perform a complete circle, the blades in the downstream part of the turbine are influenced by the wake resulting from the upstream blades. The available power in water flow is given byPower coefficient is then defined by the aspect ratio of Savonius bucket is defined by the relation as follows [6]:

The experiments were carried out in an irrigation canal in LubukLinggau, Indonesia. The water velocity of the free stream is constant at 0.71 m/s due to the irrigation system of overflow in the main river. Water flow is coming from river in which the flow is diverted into the irrigation canal by using a sluice gate. The canal has 2.2 m width and 0.8 m depth (Figure 4). The turbine rotor was set in the middle of the canal to avoid the effect of the boundary layer of flow near the canal wall.

Figure 4: Irrigation canal for turbine test.

RESULTS AND DISCUSSION

Figure 5 shows the influence of aspect ratio of Savonius rotors on coefficient of torque and power of Darrieus-Savonius turbine without deflector, which are compared to the solo Darrieus rotor. The torque of coefficient of solo Darrieus turbine has a peak value of 0.133 at , while the Darrieus-Savonius turbine with has a torque coefficient of 0.126 at . This combined turbine shows a lower torque coefficient than solo Darrieus rotor and when the aspect ratio decreases (smaller swept areas), the coefficient becomes much lower. On the left part of the curves, there are no data points because of the overload of turbine with higher load and the rotor stops to run.

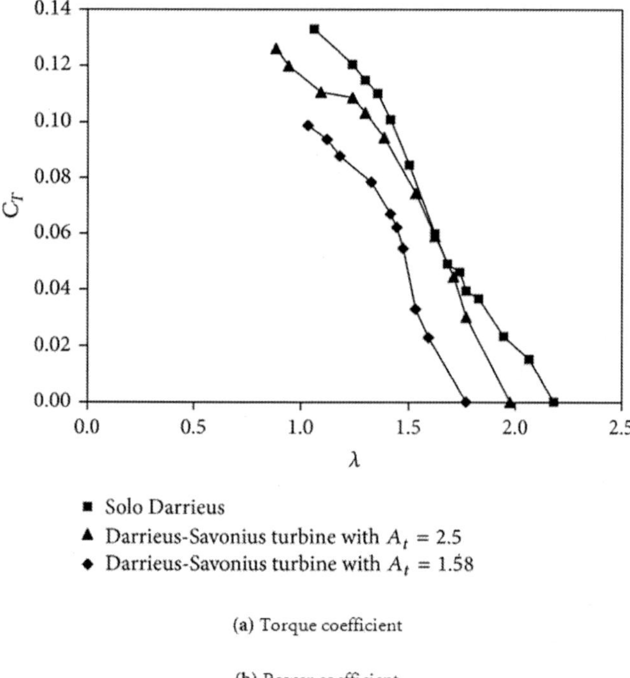

(a) Torque coefficient

(b) Power coefficient

Figure 5: Characteristics performance of Darrieus-Savonius turbine without deflector.

The performance of Darrieus-Savonius rotor is also shown by Figure 5(b) which presents the relation between the power coefficients and tip speed ratio. It is observed that solo Darrieus rotor has peak power coefficient of 0.149 when compared to the Darrieus-Savonius rotor. For combined turbine with (smaller frontal area of the bucket), the peak power coefficient is 0.134 and for a smaller aspect ratio, the coefficient is much lower. Both types of the combined turbines have a lower coefficient of power than the solo Darrieus. It is obvious that the bigger the Savonius bucket in combined turbine, the smaller the characteristics of performance, and the bigger bucket makes a higher drag force and reverse force because the shear stress becomes considerably significant. The convex surface side of the bucket may experience high skin friction drag and the concave side may experience the induced drag force. These parameters cause the deceleration of turbine rotation. The profile of torque and power coefficient shifts to the left part which shows that the rotation of Darrieus-Savonius turbine is smaller than the solo Darrieus turbine.

The performances of the turbines with deflector are presented in Figure 6. The torque profiles are similar to the torque of turbine without deflector (Figure 5(a)) in which the solo Darrieus turbine has a higher torque for high rotation. We observe that using the deflector, all turbines have higher torque compared to turbines without deflector. At peak condition, the increase is considerably significant and it is about 30% for solo Darrieus and 40% for Darrieus-Savonius turbine with . It seems that at a lower tip speed ratio, the torque of Darrieus-Savonius rotor with is higher than that of solo Darrieus. There is a little improvement of turbine torque at a lower speed.

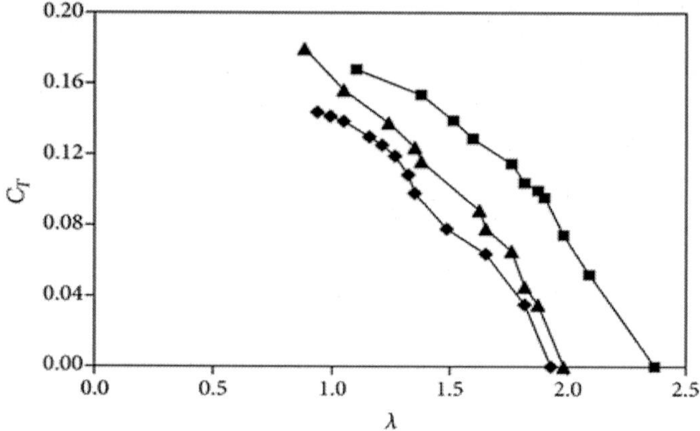

(a) Torque coefficient

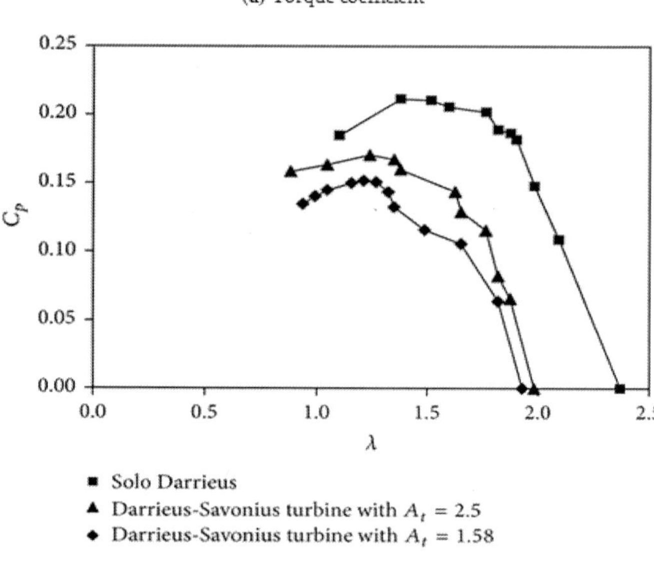

(b) Power coefficient

Figure 6: Characteristic performance of Darrieus-Savonius turbine with deflector.

The increase of the torque coefficient is followed by the increase of the power coefficient as shown by Figure 6(b). At condition of peak value, the power coefficient increases. For solo Darrieus turbine, the increase achieves 41% and for Darrieus-Savonius turbine with , the increase is about 30% or the maximum value increases from 0.149 to 0.211 for solo Darrieus turbine and from 0.134 to 0.171 for Darrieus-Savonius turbine with .

We observe further, the presence of the deflector has a very little influence upon the tip speed ratio of Darrieus-Savonius and solo Darrieus turbine as shown by Figures 5 and 6. In fact, the number of rotations of turbine with deflector increased but when it is transformed into the dimensionless tip speed ratio. It becomes little changes in variation of . By setting the deflector at a certain angle, the free stream velocity increased due to concentration of flow and this makes the increase of turbine rotation so that it may be constant as stated by the definition of the tip speed ratio. The presence of deflector, the coefficient of torque, and coefficient of power increase significantly as shown by Figures 5 and 6.

We observe all configurations of Darrieus-Savonius turbine containing Savonius rotor and that the rotational speed becomes lower than solo Darrieus rotor. The characteristic performances decrease from the solo Darrieus performance. It may be caused by geometric of a bucket of Savonius rotor. The higher bucket dimensions decrease the rotor performance and it is due to higher skin friction drag of viscous flow water on the outer surface of the convex bucket in all complete rotations of the rotor. The other part, the concave side of bucket section, produces induced drag when rotating at returning side.

CONCLUSIONS

From the above discussions, it is concluded that installing of Savonius in Darrieus rotor without deflector makes the Darrieus-Savonius rotor have lower performance characteristics than solo Darrieus rotor, but the torque has a little increase. By setting a deflector plate in upstream rotor of Darrieus-Savonius turbine, the power coefficient characteristics and torque coefficient characteristics increase significantly. In Darrieus-Savonius, the aspect ratio of Savonius buckets affects the turbine performances because a higher performance is obtained for a higher aspect ratio.

ABBREVIATIONS

A_t : Aspect ratio
B: Number of blades
C: Chord length of blade (m)
C_p : Power coefficient
C_T : Torque coefficient
d_p : Diameter of brake wheel (m)
d_D : Diameter of Darrieus rotor (m)

d_r : Rope diameter (m)
d_S : Diameter of semicircular bucket (m)
D_S : Total diameter of Savonius rotor (= $2d_S$)
F_S : Load on spring scale (kgf)
F_W : Load on weighing pan (kgf)
g: Gravitation force (m/s^2)
h: Height of deflector position (m)
L_D : Length of Darrieus blade or rotor (m)
L_S : Length of Savonius bucket or rotor (m)
P: Water power (Watt)
P_S : Shaft power (Watt)
R: Radius of Darrieus rotor (m)
T: Torque (Nm)
U: Tangential velocity of Darrieus rotor (m/s)
V: Free stream velocity (m/s).

CONFLICT OF INTERESTS

The authors declare that there is no conflict of interests regarding the publication of this paper.

ACKNOWLEDGMENT

This paper is supported by the BOPTN Fund of Sriwijaya University.

REFERENCES

1. A. N. Gorban', A. M. Gorlov, and V. M. Silantyev, "Limits of the turbine efficiency for free fluid flow," Journal of Energy Resources Technology, vol. 123, no. 2-4, pp. 311–317, 2001. View at Scopus
2. J. D. Winchester, S. D. Quayle, and S. D. ", "Torque ripple and variable blade force: a comparison of darrieus and gorlov-type turbines for tidal stream energy conversion," in Proceedings of the 8th European Wave and Tidal Energy Conference, Uppsala, Sweden, 2009.
3. M. Shiono, K. Suzuki, and S. Kiho, "Output characteristics of darrieus water turbine with helical blades for tidal current generations," in Proceedings of the 12th International Offshore and Polar Engineering Conference, pp. 859–864, Kitakyushu, Japan, May 2002. View at Scopus
4. J.-L. Menet, "A double-step Savonius rotor for local production of electricity: a design study,"Renewable Energy, vol. 29, no. 11, pp. 1843–1862, 2004. View at Publisher · View at Google Scholar · View at Scopus
5. D. Matsushita, K. Okuma, S. Watanabe, and S. Furukawa, "Simplified structure of ducted Darrieus type hydro turbine with narrow intake for extra low head hydropower utilization," Journal of Fluid Science and Technology, vol. 3, no. 3, 2008.
6. M. J. Alam and M. T. Iqbal, "A low cut-in speed marine current turbine," Journal of Ocean Technology, vol. 5, no. 4, pp. 49–62, 2010. View at Scopus
7. M. N. I. Khan, M. Tariq Iqbal, M. Hinchey, and V. Masek, "Performance of savonius rotor as a water current turbine," Journal of Ocean

Technology, vol. 4, no. 2, pp. 71–83, 2009. View at Scopus

8. K. Golecha, T. I. Eldho, and S. V. Prabhu, "Investigation on the performance of a modified savonius water turbine with single and two deflector plates," in Proceedings of the 11th Asian International Conference on Fluid Machinery, IIT Madras, Chennai, India, November 2011.
9. Y. Kyozuka, "An experimental study on the Darrieus-Savonius turbine for the tidal current power generation," Journal of Fluid Science and Technology, vol. 3, no. 2, 2008.
10. S. Kaprawi, D. Santoso, and A. Radentan, "Performance of combined water turbine with semielliptic section of the savonius rotor," International Journal of Rotating Machinery, vol. 2013, Article ID 985943, 5 pages, 2013. View at Publisher · View at Google Scholar
11. M. Takao, H. Kuma, T. Maeda, Y. Kamada, M. Oki, and A. And Minoda, "A straight-bladed vertical axis wind turbine with a directed guide vane row—effect of guide vane geometry on the performance," Journal of Thermal Science, vol. 18, no. 1, pp. 54–57, 2009.

Chapter 3

HIGH SPEED ROTORS ON GAS BEARINGS: DESIGN AND EXPERIMENTAL CHARACTERIZATION

G. Belforte[1], F. Colombo[1], T. Raparelli[1], A. Trivella[1] and V. Viktorov[1]

[1]Department of Mechanical and Aerospace Engineering, Politecnico di Torino, Italy

INTRODUCTION

Gas bearings are employed in a variety of applications from micro systems to large turbo-machinery. As they are free from contaminants if supplied with clean air, gas bearings and pneumatic guide-ways are often used in food processing, textile and pharmaceutical industries. The new research works are focused on expanding the applications of gas bearings, in particular at very high speeds. Dental drills for example operate at speeds of over 500 krpm and it seems that a limit for gas bearings without cooling is 700 krpm [1]. Nevertheless in [2] a spindle with 6 mm diameter that operated at 1.2 million rpm is described.

Because of the extremely close manufacturing tolerances that air bearings require and the lack of standard large scale production models, their costs are not at all competitive with those of the rolling bearings in common use. In order to determine whether the initial costs associated with investing in gas bearings will result in savings, each type of technology should be carefully examined. The service life of gas bearings is in fact practically unlimited, since they require almost no maintenance and do not wear.

Many investigations of air bearings have been conducted using experimental, numerical and theoretical approaches with analytical models, e.g. [3-6]. However research is still necessary to improve stiffness, load capacity and stability. At present, research studies potential designs individually to seek the main requirements for a particular application. For dynamic gas bearings, applications are currently limited to those involving low power, though an increasing amount of work is focusing on developing reliable solutions for higher-power uses. Machine tool applications, for example, require a stiffness comparable to those of the rolling bearings in common use; in very high speed applications operational stability is essential. In many cases, parameters such as the number and diameter of supply holes, their arrangement, and supply system geometry come into play. Where rotor stability under low load at very high rotational speed is the prime consideration, designs which bring rotor orbit amplitude down to acceptable levels can be adopted.

The design of gas bearings involves matching the load and stiffness requirements with bearing clearance, orifice type, flow rate and air supply pressure. Numerical calculations can assist bearing design, but their validity must be verified through basic experimental investigations. Therefore at the Mechanical and Aerospace Engineering Department of Politecnico di Torino both experimental and numerical methods were used to design gas bearing spindles and other rotors.

This chapter provides an overview on the design of rotor-gas bearing systems and the experimental activity carried out. For each application developed it is also presented the state of the art that can be found in literature. The models developed to simulate the rotor-bearings systems are described in a separate paragraph.

Four prototypes of high speed spindles were designed using gas bearings: a completely pneumatic spindle, an electro-spindle designed for machine tools, a rotor for textile applications and a mesoscopic spindle devoted to high precision machining of micro-parts at very high speeds.

THE PNEUMATIC SPINDLE

In high speed machining there are some applications for drilling, milling, and grinding, in which gas bearings are used to support the spindle [7,8]. The spindle technology in ultra-precision turning and grinding is nowadays an integration of the motor, spindle shaft and the bearings. In general these spindles have diameters smaller than 20 mm and it is difficult to find an application with a pneumatic spindle of greater diameter. In reference [9] a prototype for woodworking with spindle diameter 60 mm is described.

The prototype developed at Politecnico di Torino is capable of achieving 100000 revolutions per minute and operates at an air supply gauge pressure of 0.4-0.6 MPa. It was designed with the purpose of obtaining high load capacity and stiffness on bearings, so the spindle diameter is greater than spindles designed to achieve 200000 rpm.

The spindle is shown in Figure 1, which illustrates how the housing (4) is constrained to the base through flange (5) and journal (6). Radial support is provided by bushings (7) and (8), while axial thrust is opposed by disks (9)-(11).

The housing is made of 18 Ni Cr Mo 5 steel, while the rotor (Figure 2) (mass 7 kg, diameter 50 mm, length 459 mm) is made of 88 Mn V 8 Ku tool steel quenched and tempered to a hardness of 60 HRC. The rotor was also aged in liquid nitrogen for 5 h and dynamically balanced to a grade better than ISO quality grade G-2.5. The nose (12) to which loads are applied is secured to one end of the rotor, while the driving turbine (13) is integral with the other end. Bushings are made of the same material as the housing and have an axial length of 100 mm.

The bearings were designed to maximize the stiffness because of the importance of this parameter during cutting operations. They are provided with four circumferential sets of four 0.25±0.01 mm diameter radial holes, drilled in brass inserts as shown in Figure 3.

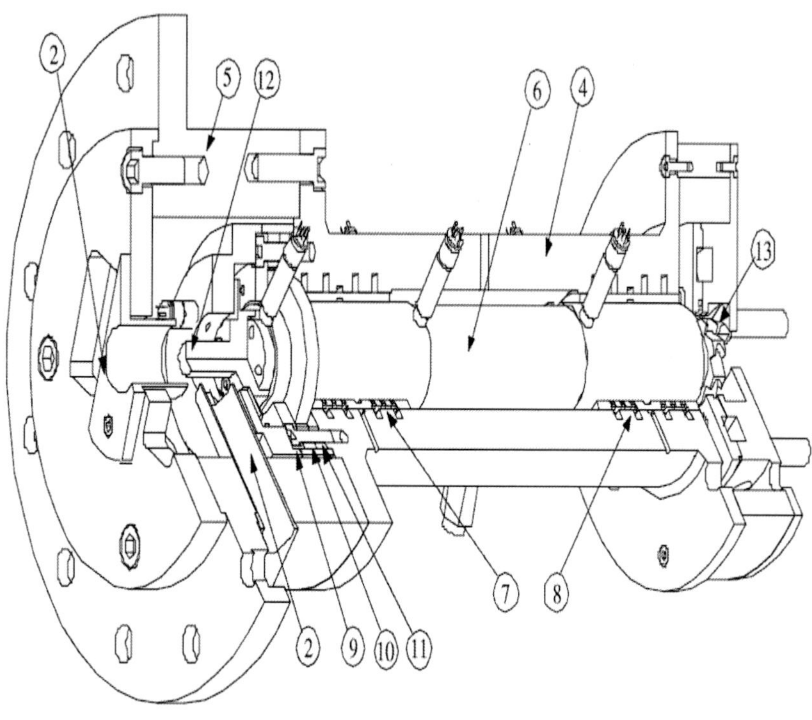

Figure 1. Section of the pneumatic spindle

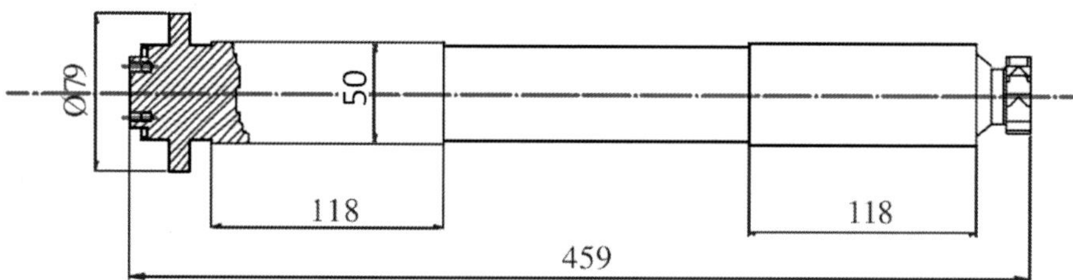

Figure 2. Rotor

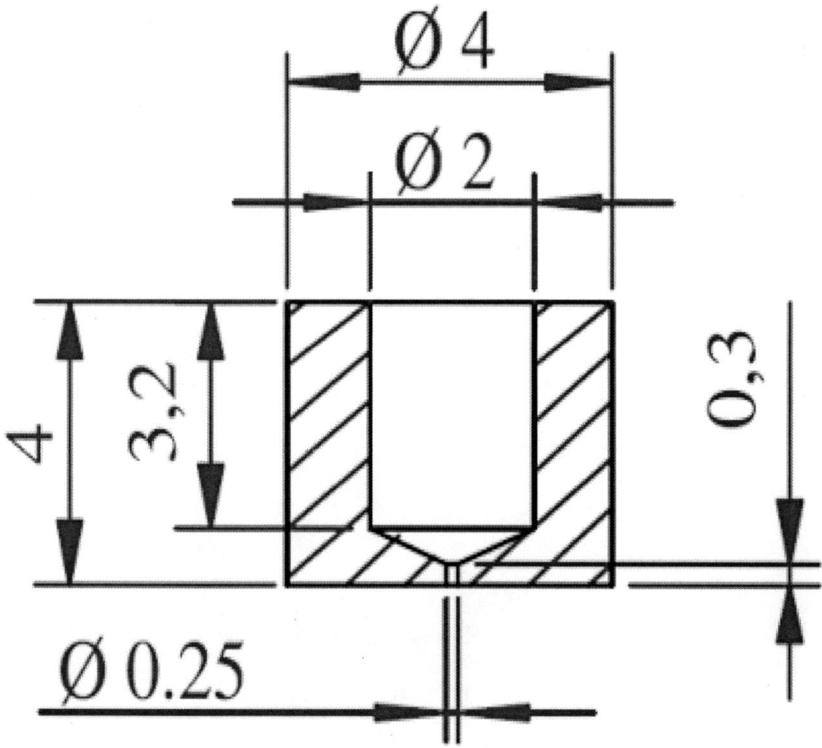

Figure 3. Brass insert with the supply hole

The axial thrust (Figure 4) is controlled by two disks (9) and (11) facing the flange on the journal. These disks are separated by a ring (10) whose thickness determines the size of the air gap. Supply air is delivered from an axial hole in the housing, is distributed through a circumferential slot, and then crosses a series of axial and radial channels machined in the disks to reach 0.25±0.01 mm diameter axial nozzles (14) and (15), which are also machined in inserts. Both the bushings and the disks were surface hardened and machined to produce a surface roughness of 0.2 and 0.4 μm, respectively at the air gaps.

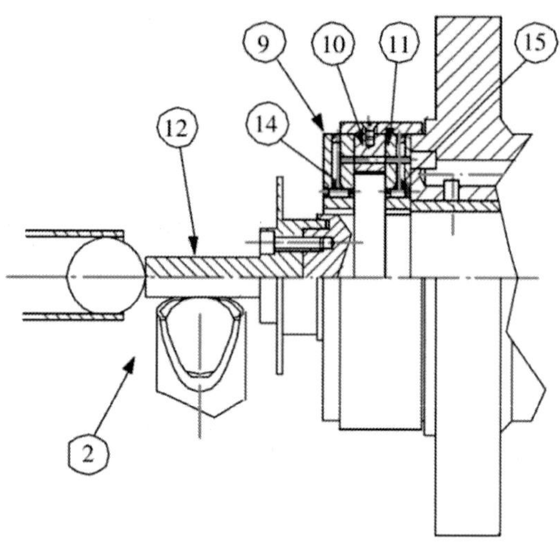

Figure 4. Enlargement of the thrust bearing and the nose

Radial and axial forces are applied to nose (12) by means of load devices (2). These devices are made of a hollow cylinder containing a calibrated sphere with a diametral clearance of 40 μm. When the cylinder chamber is supplied, the sphere is pushed against the nose and at the same time supported, so that it can rotate against the nose without sliding. Radial and axial forces can thus be transmitted to the rotor even when the latter is in motion.

Supply air for the turbine (Figure 5) crosses pre-distributor (16) in the axial direction to reach annular chamber (17), from which distributor (18) leads to eight tangential channels. Air is exhausted after actuating the turbine. Open loop speed control is accomplished by establishing turbine supply pressure.

Bearing supply is separate from turbine supply. Should the air supply fail, a reservoir enables the bearings to operate during rotor deceleration, thus preventing the rotor from seizing on the bushings or disks. Supply lines are provided with two air filtration units featuring borosilicate glass microfiber cartridges whose filtration efficiency is 93 and 99.99% respectively with 0.1 μm diameter particles. For the bearing supply line, an activated carbon coalescent filter was added to eliminate any oil vapors.

The test bench uses five capacitive displacement transducers with 0.1μm resolution, 500 μm full scale reading and 6 kHz passband. One of the transducers is used axially to measure the relative position of the rotor and thrust disks. The other four are installed radially on two different planes at right angles to the rotational axis. Rotor displacement in the bushing can thus be measured in both plane directions. The signals from these sensors are amplified by appropriate charge preamplifiers and then conditioned in a module containing a single oscillator and a demodulator for each channel. The sensitivity of these sensors is constant within the 0-10 V linearity range.

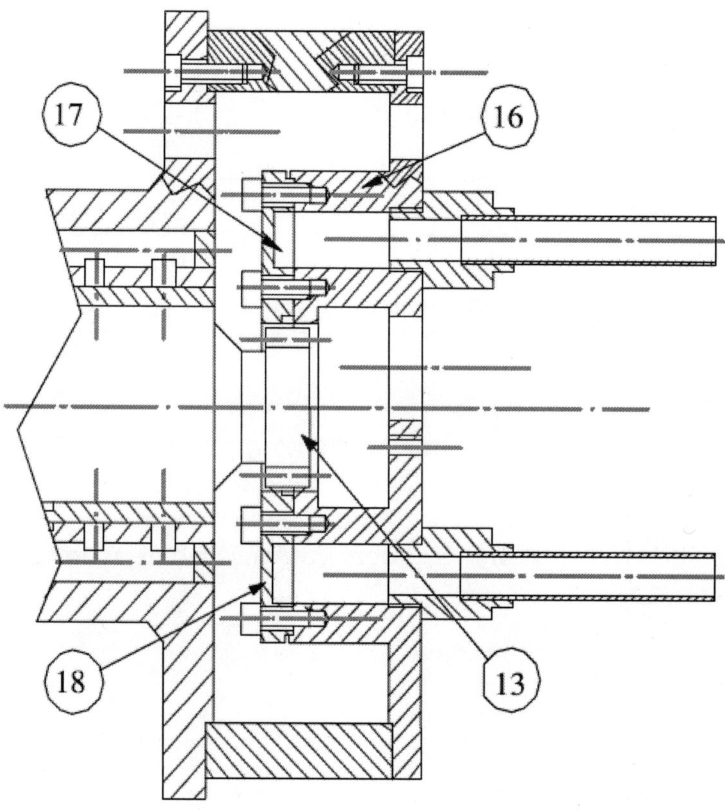

Figure 5. Enlargement of the driving turbine

To determine the journal rotational speed, an optical tachometer provided with an emitter and receiver is used, with a digital counter. Several thermocouples are also used to measure the temperature of the outer housing surface and of the air issuing from the bearing exhaust ports.

Dimensional checks were carried out to know with good precision the air gap of the bearings and the diameters of the holes. The mean inside diameter of the bushing and the mean external diameter of the rotor were measured with a precision height gage (Mitutoyo Linear Height). The mean radial air gap was calculated as the difference between the radius of the bushing and the radius of the rotor. The axial mean air gap between thrust flange and disks is calculated as the difference between measured ring and flange mean thicknesses. Results are compared with the nominal air gaps in Table 1.

Table 1. Measured values of the air gaps

	Radial air gap (μm)	Axial air gap (μm)
exp. value	26	18
nominal value	25	20

The diameters of the holes that supply the bearings were checked with an optical fiber camera with 50x and 100x magnifying lenses. Table 2 shows the measured mean diameters of the holes, with the indication of the frequency. These holes were produced using microdrills.

Table 2. Measured diameters of the supply holes

Diameter (mm)	Frequency
0.24	25
0.245	7
0.25	16

Tests were carried out to determine the bearing stiffness with the rotor stationary. The radial stiffness measured in correspondence of the nose at 120 mm from the front side of the bearing is 18 N/μm at 0.6 MPa supply gauge pressure. The axial stiffness is 27 N/μm at the same supply pressure.

Figure 6 shows the thermal transient at 40 krpm for spindle internal and external temperature measurements. The internal temperature is close to that of the air issuing from the exhaust ports. This is in accordance with the results indicated in the literature, see e.g. [10].

Rotor orbits at the two radial bushings were recorded at speeds up to 50 krpm, although tests have gone up to 80 krpm.

Figure 7 shows an example of orbits at 45 krpm, both in forward precession. Sensors 1 and 2 are for the bushing on the turbine side, while 3 and 4 are for the bushing on the motor side. These orbits, which were measured with zero radial and axial loads, are synchronous and stable. As signal frequency analysis indicated that no peak appears at a frequency of around half the rotation frequency, unstable whirling does not occur.

The centrifugal forces effect has been taken into account during the rotor designing. The radial deformation of the rotor far from its flange is visible in Figure 8.

The approaching of the external surface of the rotor to the sensors has also been considered in order to plot the orbits.

By means of a finite element code a circumferential groove was designed in proximity of the rotor flange in order to compensate the deformation due to the centrifugal force of the flange.

In Figure 9 the calculated deformation with the circular groove (depth 0.5 mm, length 10 mm) is visible. The deformation is enlarged with respect to the rotor profile. Also thermal effects on the relative distance between rotor and sensors, mounted on the housing, have been taken into account in order to individuate the centre of the orbit.

THE ELECTRO-SPINDLE

In literature can be found examples of air bearing electro-spindles for high speed and high precision applications, see for example references [11-13]. The electro-spindle developed at Politecnico di Torino [14], shown in Figure 10, is composed of a rotor of 7 kg mass, 50 mm diameter and 479 mm length. It is supported by air bearings and accelerated by means of an asynchronous motor mounted on one end of the spindle. On the opposite end of the rotor a clamping tool is mounted.

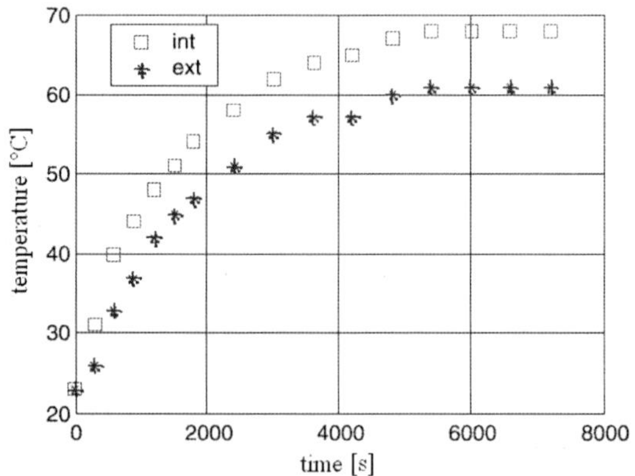

Figure 6. Thermal transient

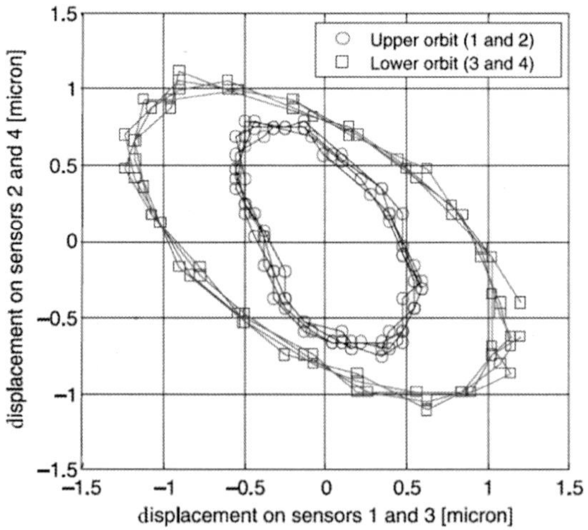

Figure 7. Rotor orbits = 45 krpm

Figure 11 shows a section of the electro-spindle with carter (1), rotor (2), two bushings (3) and double thrust bearing (4). Motor (5) is of the two-pole squirrel-cage type controlled by an inverter. Speed range is up to 75 krpm and power is 2.5 kW. A clamping tool designed for high-speed is screwed onto the left end of the rotor. By mounting a tool on the spindle it is possible to test the dynamic behavior of the rotor also during the machining process.

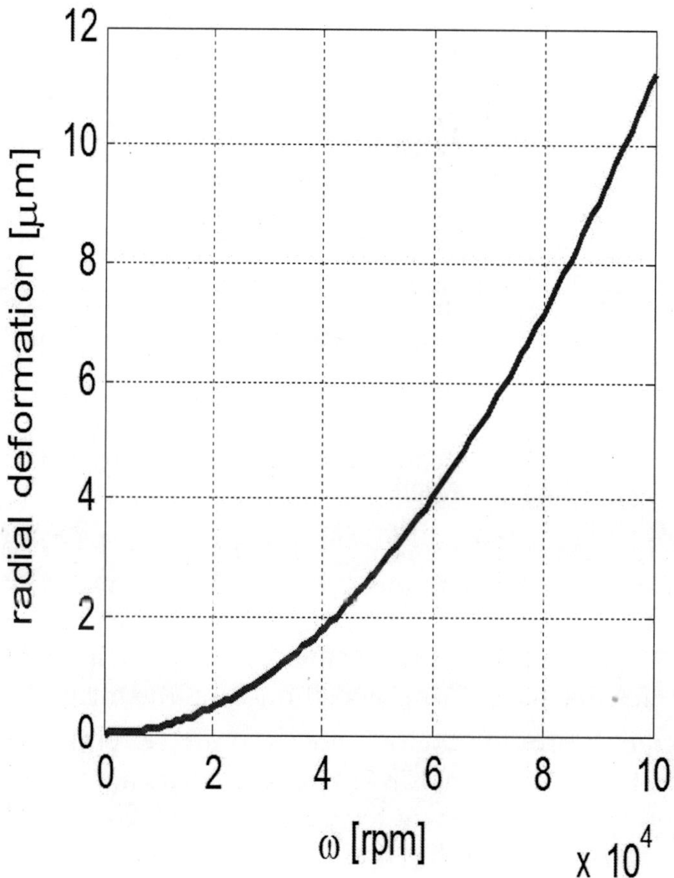

Figure 8. Centrifugal expansion of the rotor in correspondence of the bushings (diameter 50 mm)

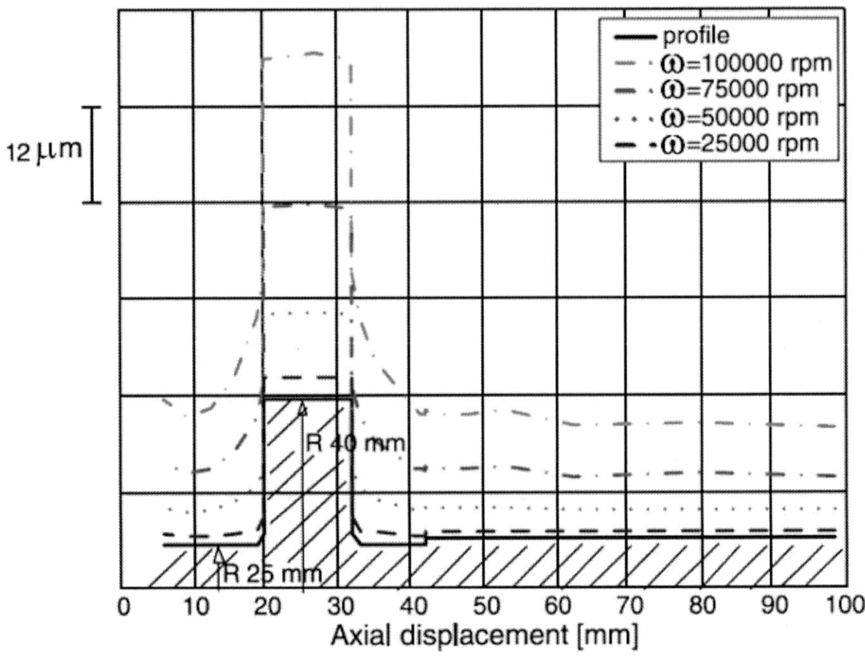

Figure 9. Rotor deformation due to centrifugal force in correspondence of the flange at different rotational speeds

The radial bearings feature cylindrical barrels. Each barrel has four sets of supply ports diameter 0.25±0.01 mm arranged 90 degrees apart. The thrust bearing, similar to the one described in [15], is composed of two disks facing the flange on the journal. Each disk has 8 axial nozzles dia. 0.2±0.01 mm positioned on the mean diameter.

The system is provided with a closed cooling circuit that controls the temperature of the motor and of the discharge air. Without refrigeration and with ambient temperature 298 °K, at 60000 rpm the temperature would reach 383 °K after two hours due to power losses on bearings.

An optical tachometer facing the rotor was provided to measure rotational speed. Four capacitance displacement transducers were inserted radially and at right angles in the carter facing the rotor to measure dynamic runout. Two were positioned on motor side, the other two on thrust bearing side.

Clearances were measured moving the rotor axially and radially until contact is made. It was found that axial and radial clearances are about 15 μm and 20 μm respectively.

In order to measure radial and axial stiffness of the tool, the electro-spindle is mounted on a test rig designed for the purpose with proper load devices. Radial forces were measured at different supply pressures (Figure 12) at=0. Figure 13 shows the axial load capacity readings for 0.3, 0.5 and 0.7 MPa supply absolute pressure. The load device was used for positive displacements and weights were applied to obtain the curve with negative displacements.

The rotor orbits depicted in Figure 14 were measured at the same supply pressure. Due to the rotor centrifugal expansion these orbits appear not to be centered in the bushings because the relative rotor-sensor distance decreases. The spindle was tested up to 53000 rpm and the tests were stopped because of the high rotor vibration. The permissible residual imbalance should be diminished in order to allow tests at higher speeds. Anyway the whirl instability did not occur and the imbalance response was only synchronous.

Figure 10. Photo of the electro-spindle

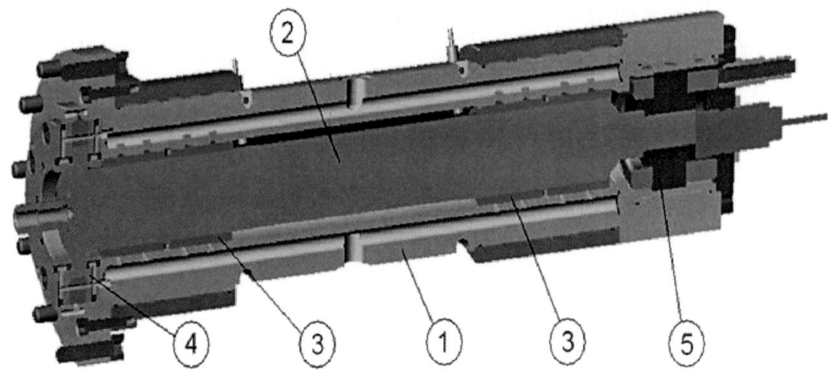

Figure 11. Schematic section of the electro-spindle

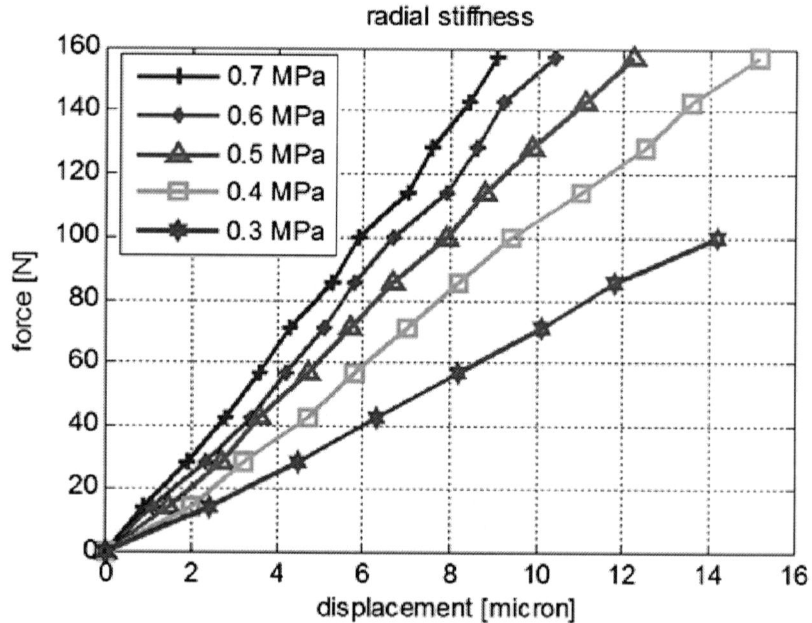

Figure 12. Radial force on the tool (measured at 120 mm from the front side of the bearing) versus radial displacement at different bearing supply absolute pressures

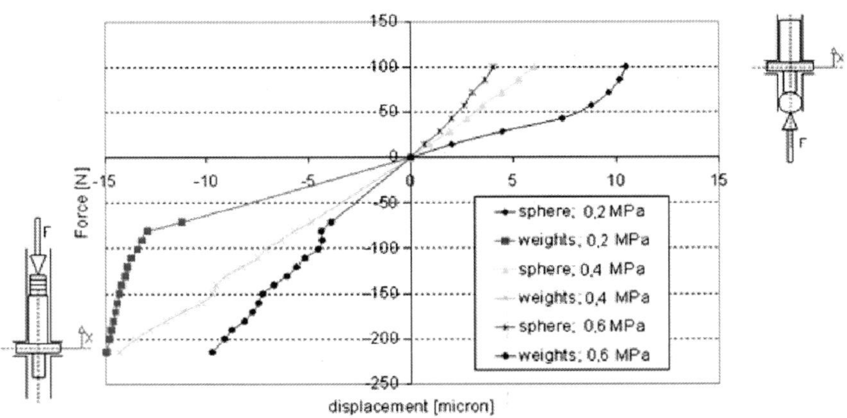

Figure 13. Diagram of axial force on tool versus displacement at different bearing supply gauge pressures

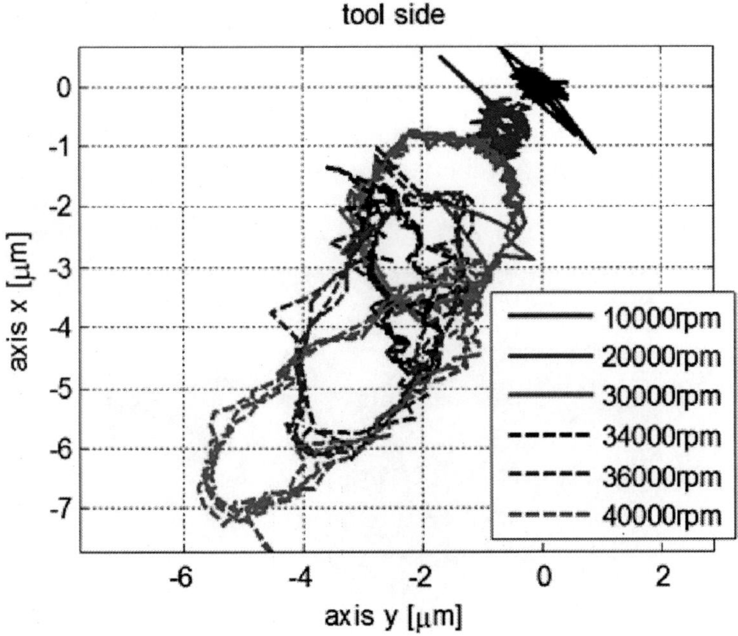

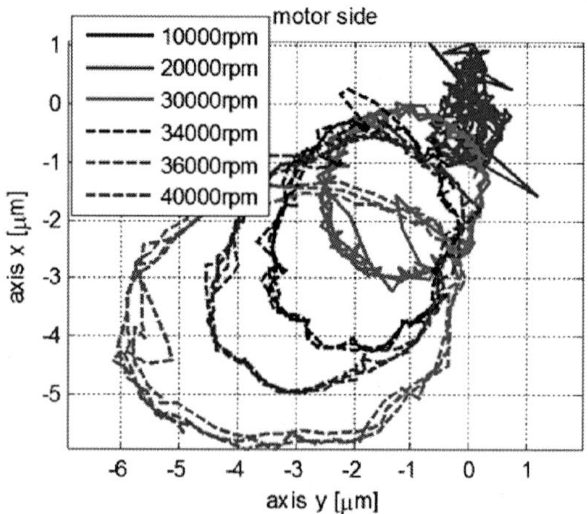

Figure 14. Rotor orbits in correspondence of a bushing due to the residual unbalance; supply gauge pressure 0.6 MPa

The electro-spindle was also tested in dynamic conditions during machining with high speed milling cutters of diameters in the range 1 to 6 mm. The system depicted in Figure 15, mounted below the electro-spindle, provides the advance along axis x of the material under milling. The material under machining was a block of rapid prototyping resin, advanced by means of a motorized slide. The tests were made up to 40000 rpm with feed speeds from 1 to 10 mm/s and chip thickness 1 mm.

Figure 15. Motorized slide used for the dynamic tests

THE TEXTILE ROTOR WITH DAMPING SUPPORTS

Gas bearings suffer from instability problems at high speed. A method to increase the stability threshold (the speed at which the unstable whirl occurs) is to increase the damping of the rotor-bearings system by introducing external damping supports [16]. A design guideline for the selection of the support parameters that insure stability in an aerodynamic journal bearing with damped and flexible support is given in paper [2].

The prototype described in this paragraph was designed with the priority of increasing the stability at high speeds [17]. The method adopted for this purpose was the use of rubber O-rings.

The prototype consists on a rotor (1) made of hardened 32CrMo4 steel with mass 0.96 kg, diameter 37 mm and length 160 mm. The rotor is supported by a radial air bearing mounted on rubber O-rings and an axial thrust bearing (Figure 16). It was designed to rotate in stable conditions up to 150 krpm. At one end of the rotor an air turbine (2) was machined and at the other end a nose (3) was screwed to the rotor. The housing (4) is fixed to the base and has four circumferential slots in which the O-rings are inserted. The bushing (5) incorporates the rubber rings and has four sets of supply nozzles (diameter 0.2±0.01 mm) fabricated by EDM. The total length of the bearings is 57 mm. In the middle plane of the bushing a discharge slot (6) is vented by a radial hole in the housing (see Figure 17). A central annular discharge chamber separates the radial bearings.

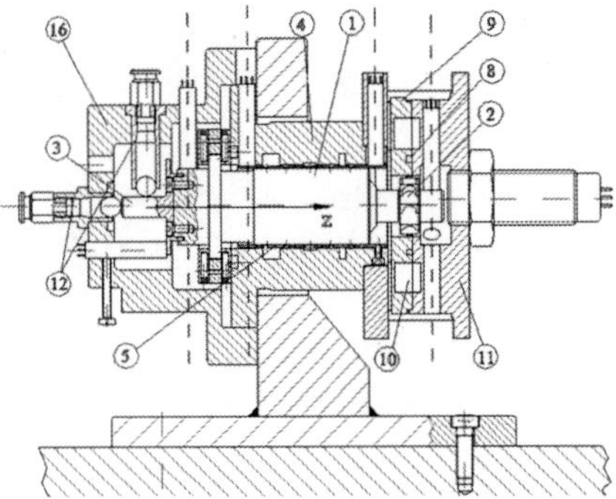

Figure 16. Test bench of the floating bushing

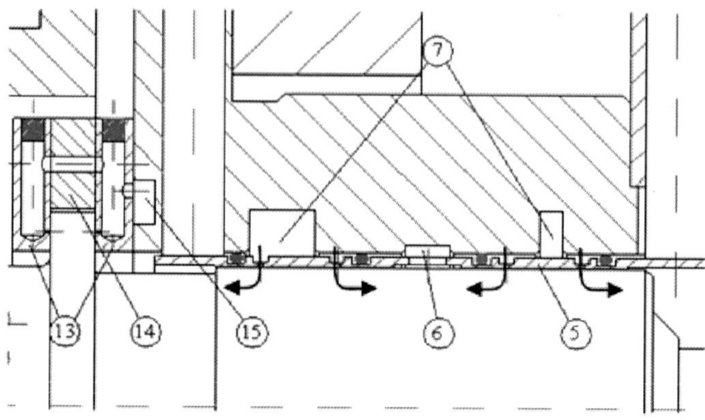

Figure 17. Enlargement of the floating bushing

The air from supply slots (7) flows to the radial clearance through the nozzles to reach the vent centrally in the discharge slot and laterally. The purpose of the O-rings, besides providing a seal between supply slots and discharge chamber, is to introduce damping in the rotor-bearing system. The turbine is driven by tangential jets discharged through 8 nozzles (8) machined on distributor (9). Annular chamber (10) connected to the nozzles is supplied through an axial hole on pre-distributor (11). Air is exhausted after actuating the turbine. Open loop speed control is maintained by setting the turbine supply pressure. The rotational speed was measured by an optical tachometer consisting of an emitter and a receiver facing the rotor at the turbine side. A retro-reflector stuck to a portion of the rotor face reflects emitted signal once per revolution.

Radial and axial forces are applied to the nose by means of loading systems (12) similar to the ones previously described. Eight capacitive displacement transducers are inserted radially in the housing, the pre-distributor and cover (16) to sense the rotor and bushing positions. An axial transducer can be inserted near the nose to monitor the axial position of the rotor with respect to the thrust bearing.

The O-Rings have 41 mm inside diameter and 70 Shore hardness. The three materials used for testing are NBR (Butadiene Acrylonitrile), Viton® (Fluorinated Hydrocarbon) and Silicone (Polysiloxane).

Accurate dimensional checks were carried out to evaluate axial and radial clearances, supply holes diameter and O-ring interference. The total diametral gap between them was found to be 35±2 μm. The difference between the thickness of central ring (14) and rotor flange was 19±2 μm, giving an axial clearance of approximately 9.5 μm.

To measure the diameter of the nozzles supplying the bearings an optical fibre camera with 200X magnifying lens was used. The measurements were accurate and repeatable, thus proving the superiority of EDM technology over micro-drilling. Figure 18 shows a sample photographic record at 200X magnification.

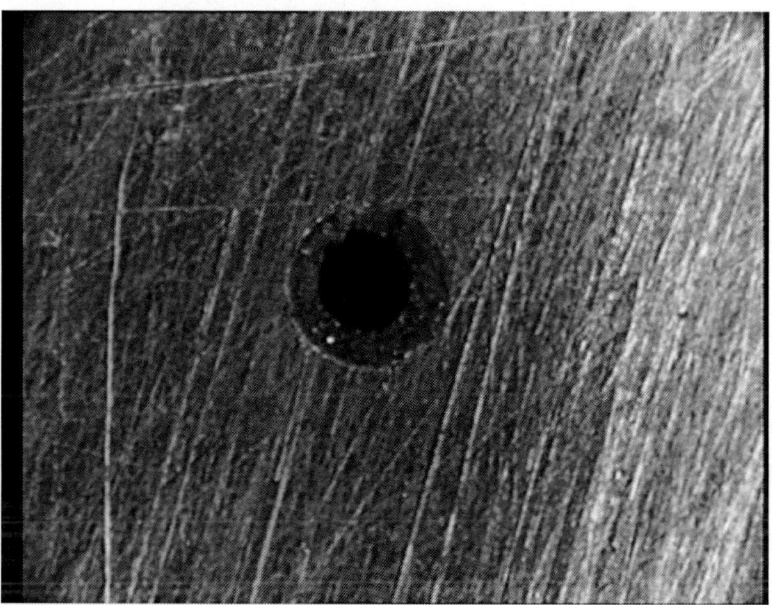

Figure 18. Supply hole magnification (200X)

The supply hole diameter, after fixing the radial clearance, is selected on the basis of numerical investigation conducted to simulate the dynamic behavior of the system. The mathematical model used for this purpose is described in a separate paragraph at the end of this chapter.

O-ring grooves in the housing have a medium diameter of 43.5 mm, while external diameter of the bushing is 41 mm. The cross section diameter *d* of the rings was determined by a shadow comparator.

Table 3 lists the interferences on the O-rings calculated using the equation

$$int\%=(d-Di-De2)100d$$

where D_i and D_e are the inside diameter of the grooves machined in the housing and the external diameter of the bushing respectively. With 0.6 MPa pressure differential the sealing function of the rubber rings between chambers 7 and 6 was realized with interference about 10% or more.

In Table 4 the measures of the inner diameter d and the cross-section diameter d_c are shown.

Table 3. Interference values

	Cross section diameter d [mm]	Interference
NBR-Silicone	1.78±0.01	30%
Viton	1.73±0.01	28%

TABLE 4. Dimensions of the O-rings

	d (mm)	d_c (mm)
NBR	41	1.80
Viton	41	1.83
Silicone	41	1.83

Measured Rubber Dynamic Stiffness

The dynamic stiffness of rubber O-rings is measured in order to introduce into the model the stiffness and the viscous equivalent damping. These parameters depend on the vibration frequency and also on the radial displacement imposed. Tests were made under different conditions, varying the diametral interference on the O-ring and the displacement amplitude x_0 imposed, in the frequency range 300÷800 Hz. In Figure 19 is visible the scheme of the test rig, in which the cylinder is fixed and the casing is mounted on the shaker plate. The force amplitude F_0 is measured by the load cell mounted between two fixed parts.

Measurements were made at different radial amplitudes. Increasing x_0 both stiffness and damping decrease. The results visible in Figure 20 were obtained with x_0=25 μm and a diametral interference of 11%, that are similar to that occur in the air bearing test bench. The results obtained with Silicone are not reported because the FRF of transfer function F/x was very noisy.

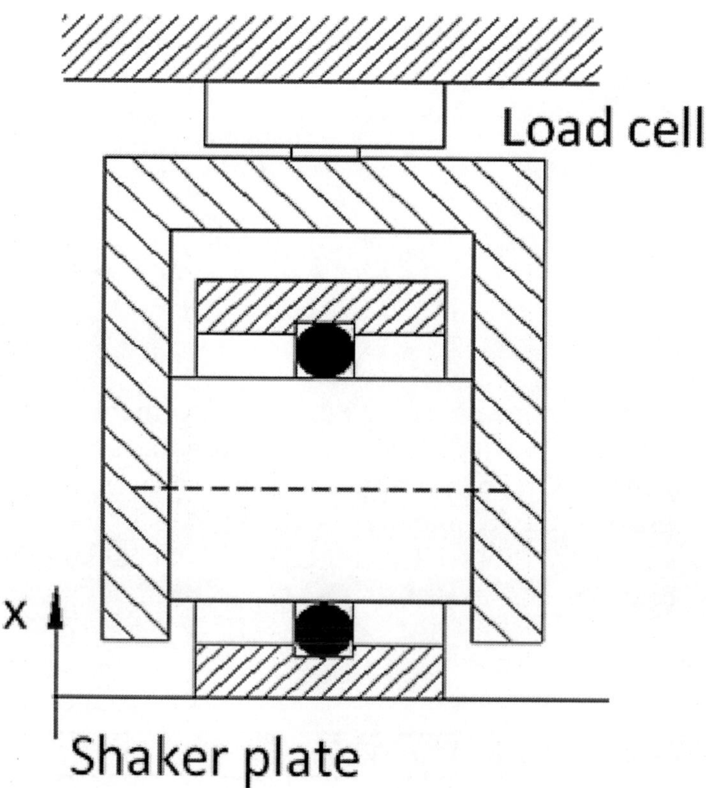

Figure 19. Scheme of the O-ring test rig

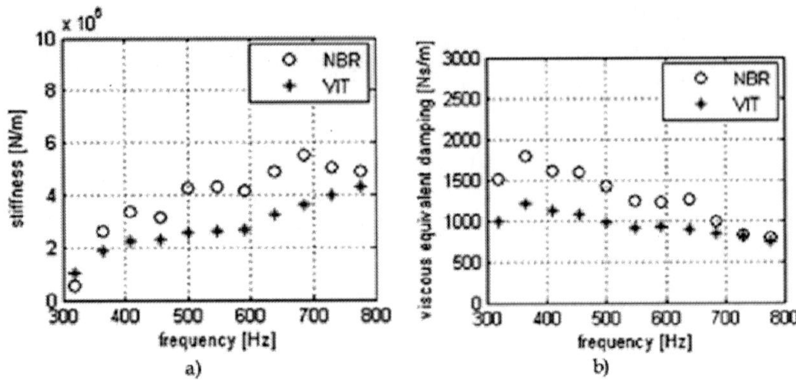

Figure 20. O-ring radial stiffness (a) and damping (b)

STABILITY

The response to a rotor radial step jump displacement of 1 μm from coaxial position is calculated. As a first approximation, average values of O-ring stiffness kOR and damping cOR are considered, neglecting the dependence on the frequency. The parameters introduced in the model are shown in Table 5. L1 and L2 are the axial lengths of the two radial bearings.

Table 5. Input values of the model

m_{rot}=0,97 kg	h_0=17 μm	L_2=23 mm	T^0=293 K
m_b=0,1 kg	d_s=0,2 mm	μ=1,81e-5 Ns/m2	k_{OR}=4·10^6 N/m
R=18.5 mm	L_1=25 mm	R^0=287 J/kgK	c_{OR}=1·10^3 Ns/m

In Figure 21 the theoretical supply pressure values in correspondence to the stability threshold are plotted vs. the rotational speed for the cases of fixed bearing and bearing mounted on O-rings. Each curve divides the plane into two regions: the upper one relative to a stable behavior of the rotor-bearing system, the lower one relative to an unstable behavior. In the first case, as a result of an initial step jump displacement of the rotor, the system evolves to the centred position (punctual stability); in the second case the rotor trajectory is an open spiral and causes the contact between the rotor and the bushing. In correspondence to the threshold curves the system evolves to a condition of orbital stability. The stabilizing effect of the rubber rings is evident because the pressure that guarantees the stability is lower.

In Figure 22 the simulated values are compared with the experimental ones, relative to three kinds of rubber: NBR, Viton and Silicone. There is good agreement between the experimental and the simulated stability threshold also if the experimental data are influenced by the rotor imbalance and in calculations the effect of imbalance is neglected (the rotor was dynamically balanced to a grade better than ISO quality grade G-2.5). The whirling frequency ς increases with the rotational speed, see Figure 23. It is interesting to observe that the whirling ratio $\vartheta=\varsigma/\mathfrak{Z}$ at the stability threshold (Figure 24) decreases with the rotational speed.

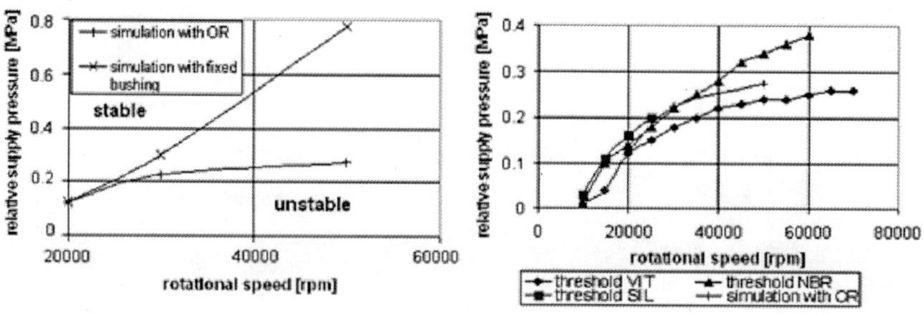

Figure 21. Theoretical results with fixed bearing and bearing mounted on OR

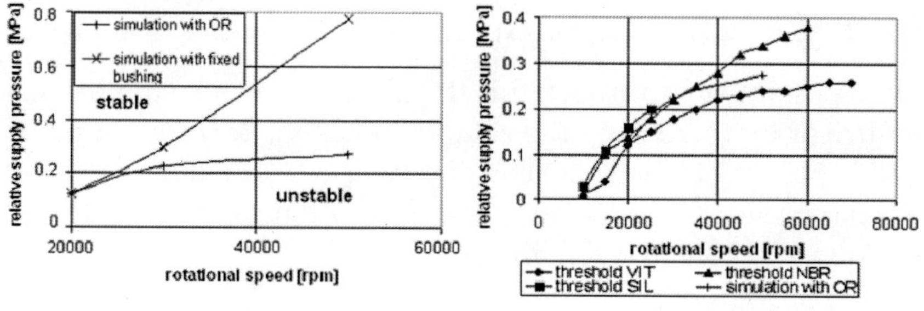

Figure 22. Comparison between experimental and simulated threshold stability

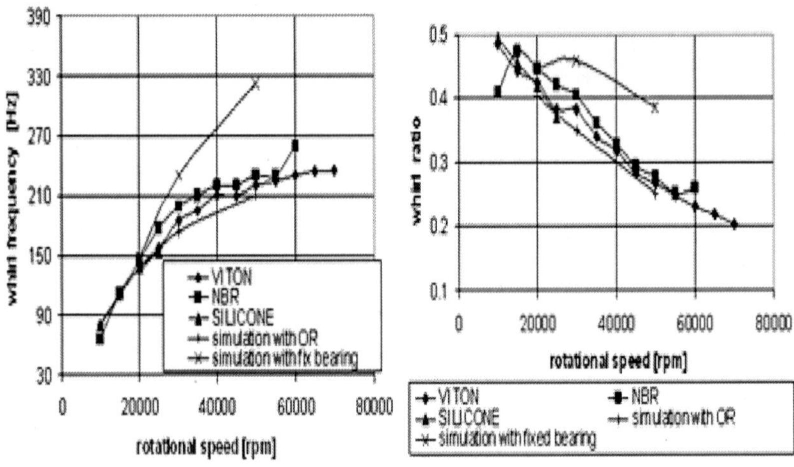

Figure 23. Comparison between experimental and simulated whirling frequency ς

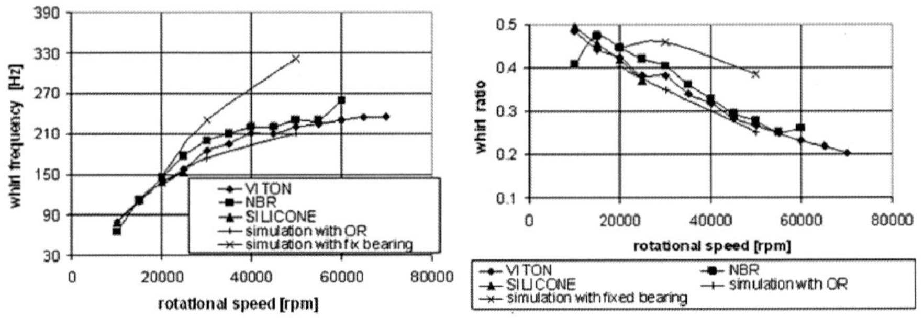

Figure 24. Comparison between experimental and simulated whirling ratio ϑ

It is possible to approach this threshold by decreasing the supply pressure or by increasing the rotational speed. Both possibilities are treated: Figures 25 and 26 show the change of the orbit amplitude vs these parameters. The increase in amplitude near stability threshold is sudden and considerable in both cases.

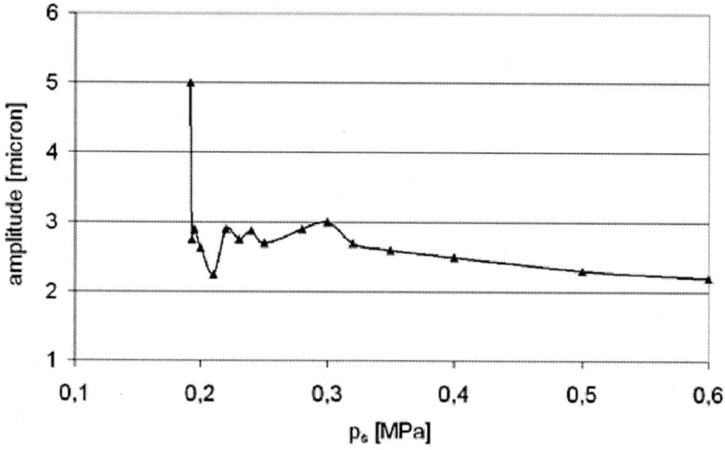

FIGURE 25. Orbit amplitude versus supply pressure; =30000 rpm

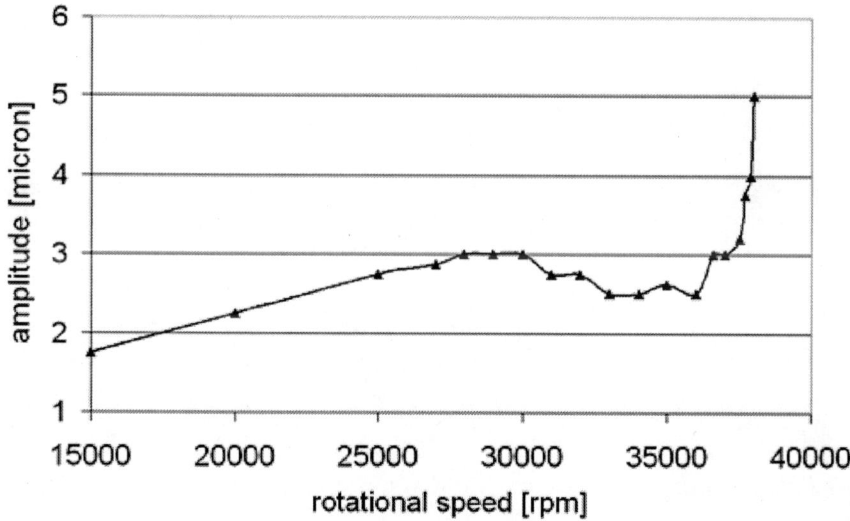

Figure 26. Orbit amplitude versus rotational speed; p_s=0.22 Mpa

Figure 27 shows the change in orbit shape with decreasing the supply pressure. Whirl motion is conical for any supply pressure at the stability threshold. Frequency spectra for turbine displacement in the two conditions are visible in Figure 28.

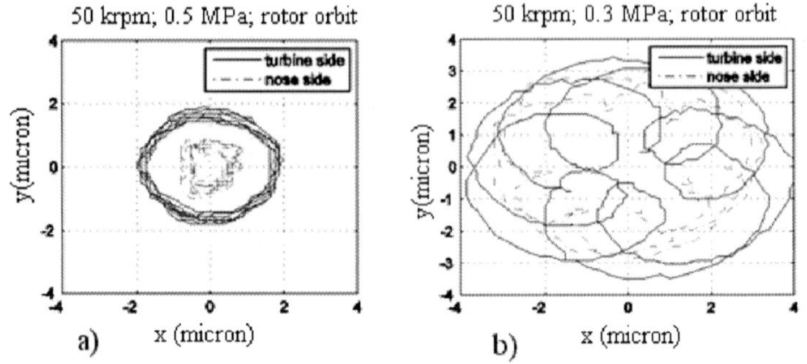

Figure 27. Rotor orbits in stable condition (a) and at stability threshold (b); =50000 rpm

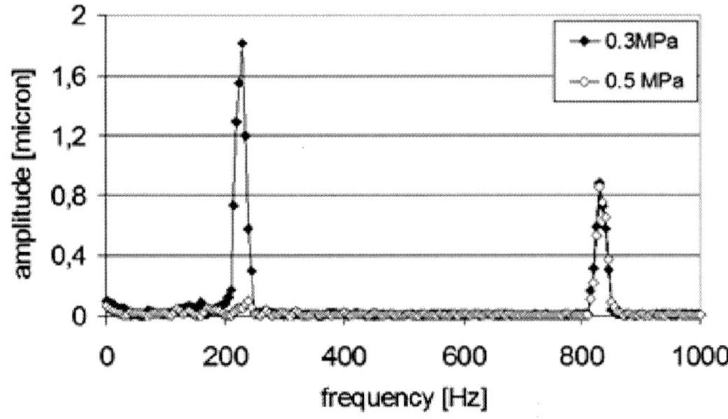

Figure 28. Frequency spectra of rotor displacement; =50 krpm

THE MESOSCOPIC SPINDLE

Another method to increase the bearing stability is to modify the film geometry from the circular journal bearing profile. Non-circular journal bearings can assume various geometries: elliptical [20-23]

offset halves [24] and three-lobe configuration [25,26] are the most common geometries. Paper [27] shows a comparative analysis of three types of hydrodynamic journal bearing configurations namely, circular, axial groove, and offset-halves.

There is an extensive literature about the study of the dynamic stability of hydrodynamic journal bearings with non-circular profile, but very few papers consider gas journal bearings of this type. The wave bearing with compressible lubricants was introduced in the early 1990's [28,29].

In the present paragraph the design of the elliptical and multi-lobes gas bearings for a ultra-high speed spindle is described [30].

The bearings were designed to have a stable regime of rotation up to 500 krpm with acceptable stiffness and load characteristics. A computerized design was used for optimization of the rotor-bearing characteristics. The bearing clearance was represented by expression

$h=h0(1+cform2(cos(n\varphi)-1))$

where c_{form} is the profile form factor, h_0 is the maximum clearance and n is the number of lobes of the profile.

The static and dynamic performances were numerically analyzed for two pairs of radial externally pressurized gas bearings. Conical and cylindrical whirl modes were considered. From numerical simulations for a 10 mm diameter rotor, bearing clearances non less than 5 μm and supply pressure 0.6 MPa the following results were obtained:

- the maximum rotor speed obtained with circular bearing (clearance 5 μm) with 4 supply orifices of 0.1 mm diameter in circumferential direction was 150 krpm, while with 32 supply orifices of 0.2 mm diameter was 250 krpm;
- with elliptical bearing profile the maximum rotor speed obtained with stable operation was 500 krpm for bearings with 4 supply orifices of 0.2 mm diameter in circumferential direction;
- the rotor with the multi-lobe bearings were less stable in comparison with the rotor with elliptical bearings;
- the positioning of supply orifices at 45° with respect to the principal axes of the elliptic profile improved bearing characteristics.

The final bearing geometry is defined by the parameters summarized in Table 6. Each elliptical journal bearing presents two rows of 4 supply orifices positioned at 45° with respect to the principal axes.

Table 6. Final bearing parameters

Maximum clearance h_0, μm	15
Rotor diameter, mm	10
Supply orifice diameter, mm	0.2
Number of supply orifices for each bearing	8
Number of bearings	4
Profile form factor $_{cform}$	0.7
Number of profile lobes n	2

Figure 29 shows the prototype of ultra-high speed spindle. The rotor, of mass 0.07 kg, is supported by two pairs radial elliptical bearings and a double thrust bearing. The calculated radial stiffness on the rotor end is 3 N/μm and the air consumption is 3.65 10^{-4} kg/s.

The axial and the radial stiffness of the bearings were measured with test benches realized at the purpose (Figure 30). The axial and radial displacement of the rotor due to an imposed load was measured by laser beams. The axial stiffness of the thrust supplied at 0.6 MPais 2.8 N/μm, while the radial stiffness is 1 N/μm. This value can be increased with a better dimensional control of the bearings internal profile.

By means of start-up (acceleration) and coast down (deceleration) tests on the spindle the bearing friction torque was estimated as a function of the speed up to 150000 rpm. The deceleration tests from different rotational speeds are depicted in Figure 31. The friction torque was found to be proportional to the rotational speed with the rate of 10^{-4} Nm every 10000 rpm. The dynamic runout of the shaft was measured by means of laser beams at different rotational speeds in correspondence of the nose.

Figure 29. Ultra-high speed spindle prototype

The unbalance response was synchronous and unstable whirl was not encountered. In Figure 32 the waterfall diagram, obtained with the FFT of the shaft radial vibration, is shown. There is a critical speed at 34000 rpm, to which corresponds a maximum spindle runout of ±9 μm.

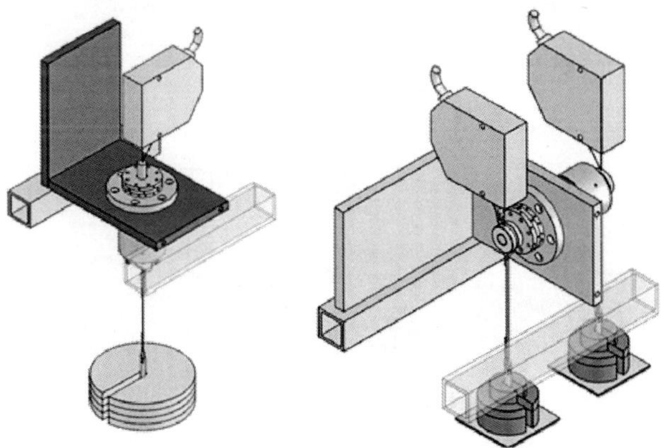

Figure 30. Test benches realized to measure the radial and axial bearing stiffness

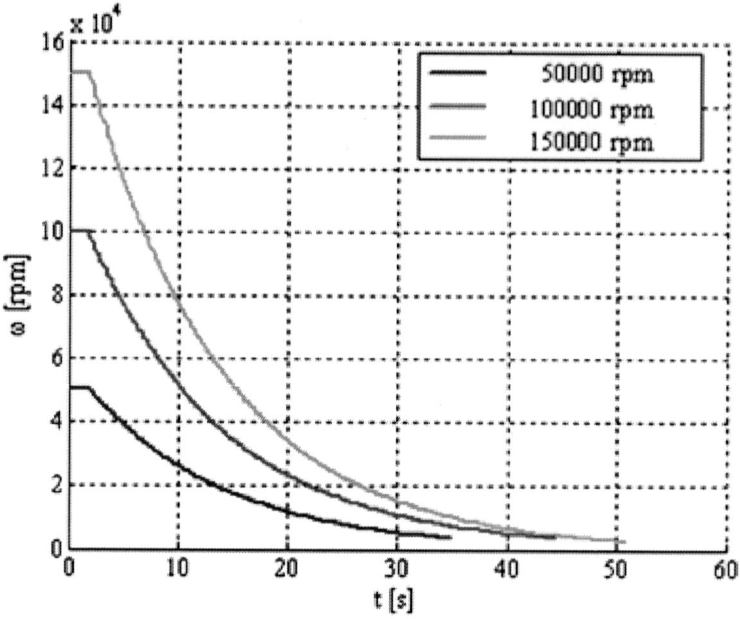

Figure 31. Test benches realized to measure the radial and axial bearing stiffness; supply gauge pressure 0.6 MPa

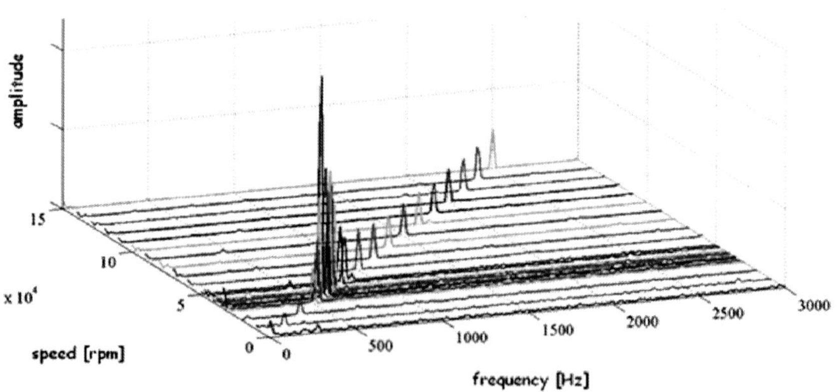

Figure 32. Waterfall diagram of the rotor unbalance response

MATHEMATICAL MODEL

The complete Reynolds equation for compressible fluid film is numerically solved together with the equations of motion of the rotor considered rigid, see reference [18].

The momentum equations for the isothermal gas lubricated films are:

pR0T0(CuCt+uCuCz+vCuRCƒ)+CpCz=−ς12uh2

pR0T0(CvCt+uCvCz+vCvRCƒ)+CpRCƒ=ς6(Rʒ−2v)h2

where *u* and *v* are the mean velocity components in *z*- and -direction (Figure 33).

They are solved together with the continuity equation

C(phu)Cz+C(phv)RCƒ+C(ph)Ct−R0T0q=0

where *q* is the inlet mass flow rate per unit surface defined by

q=GøzRøƒ

For low modified Reynolds numbers ($Re^*= h_0^2/\mu<1$) inertial terms are negligible. The equation resulting from (1-3) is simplified into the following

CCz(ph3CpCz)+CRCƒ(ph3CpRCƒ)+12ςR0T0q=6ςʒC(ph)Cƒ+12ςC(ph) Ct

Film thickness *h* is given by

h(ƒ,z)=h0−ex(z)cosƒ−ey(z)sinƒ

The rotor eccentricities e_x and e_y are related to the journal degrees of freedom by

ex(z)=xG+(z−zG)ƒy;ey(z)=yG−(z−zG)ƒx

Concerning the thrust bearing, the Reynolds equation is

r2CCr(ph3CpCr)+CCƒ(ph3CpCƒ)+12ςR0T0r2q=6ςʒr2C(ph)Cƒ+12ςr2C (ph)Ct

The boundary conditions at the discharge slots are $p=p_a$, while in correspondence of the supply ports the downstream pressure level is calculated considering orifice resistance.

The input resistance is expressed on the basis of ISO formula for flow rate through an orifice (ISO, 1989).

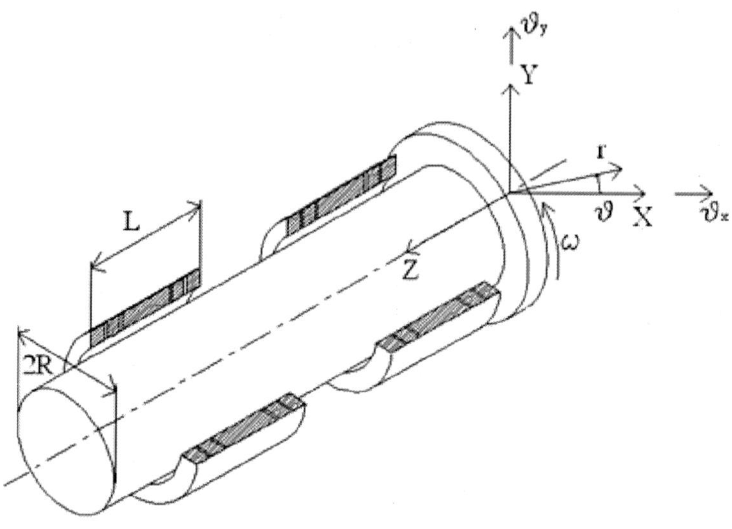

Figure 33. Schematic diagram of journal bearings

Conductance c_s appearing in the mass flow rate formula is expressed by equation (10),

cs=0.686*cdS*ϟ*NR*0*T*0√

where c_d is the discharge coefficient and *S* the cross-section of the supply orifice.

Discharge coefficient c_d depends on local clearance *h*, supply port diameter d_s and Reynolds number*Re* of supply port section. This relationship is taken into account with equation (11), see ref. [19].

cd=0.85(1−*e*−8.2*hds*)(1−*e*−0.005*Re*)

The Reynolds number is calculated with eq. (12)

Re=4*G*ϟ*ds*ς

Equation (11) is extrapolated from experimental measurements of air consumption and pressure distributions under pads for different air gap height and supply port sizes. The equation is extrapolated in the range of h/d_s values from 12.5 10^{-3} to 100 10^{-3}.

The Reynolds equation is discretized spatially by central finite-differences:

3*h*2*i*,*j*(Ϲ*h*Ϲ*z*)*i*,*jp*2*i*+1,*j*−*p*2*i*−1,*j*2∅*z*+*h*3*i*,*jp*2*i*+1,*j*−2*p*2*i*,*j*+*p*2*i*−1,*j*(

∅z)2+3h2i,j(ϹhRϹϯ)i,jp2i,j+1−p2i,j−12R∅ϯ+h3i,jp2i,j+1−2p2i,j+p2i,j−1(R∅ϯ)2+24R0T0ϛqi,j=12ϛϑhi,jpi,j+1−pi,j−12∅ϯ+12ϛϑpi,j(ϹhϹϯ)i,j+24ϛhni,jpn+1i,j−pni,j∅t+24ϛpni,jhn+1i,j−hni,j∅t

The journal equations of motion are considered together with the Reynolds equation in order to study the dynamics of the rotor-bearings system.

The 4 d.o.f. model of the rotor is described by the following system

mrẍG=Fcx+Fx+mr'ϒϑ2cos(ϑt)mrÿG=Fcy+Fy+mr'ϒϑ2sin(ϑt)JGϯẍ=Mcx+Fy(zG−zF)−JPϑϯÿ+ϥ(JP−JG)ϑ2cos(ϑt−ϥ)JGϯÿ=Mcy−Fx(zG−zF)+JPϑϯẍ+ϥ(JP−JG)ϑ2sin(ϑt−ϥ)

where J_G is the rotor transverse inertia moment, calculated with respect to the rotor centre of mass, while J_p is the polar inertia moment. The external forces have a resultant of radial components F_x, F_y applied in correspondence of $z=z_F$.

The fluid forces, comprehensive of pressure and shear actions, are expressed by

Fcx=∪L0∪20(pcosϯ−Ϣϯsinϯ)RdϯdzFcy=∪L0∪20(psinϯ+Ϣϯcosϯ)RdϯdzMcx=∪L0∪20(−psinϯ+Ϣϯcosϯ)(zG−z)RdϯdzMcy=∪L0∪20(pcosϯ+Ϣϯsinϯ)(zG−z)Rdϯdz

The tangential actions are given by:

Ϣϯ=−h2ϹpRϹϯ−ϛϑRh

The system is solved using Euler explicit method. From relation (17) the pressure for each node is calculated at iteration n+1.

pn+1i,j=pni,j+∅tf(pni,j,pni+1,j,pni−1,j,pni,j+1,pni,j−1,hni,j,hn−1i,j,(ϹhϹϯ)ni,j,(ϹhϹz)ni,j)

The explicit method has a very rapid execution per time step but is limited to the use of small Δt, see [31]. The system of nxm equations (17) is solved together with rotor equations of motion (14).

The solution procedure starts with a set of input data (shaft diameter, radial clearance, bearing axial length, position and diameter of supply orifices, shaft speed).

To calculate the static pressure distribution, h is maintained constant in time and the system is solved with initial condition $p_{i,j}=p_a$ for each node. The rotor trajectory is determined starting with the initial static pressure distribution and using the following set of initial conditions:

$x(0)=h0'Yx(0);\ y(0)=h0'Yy(0)$

$\dot{x}(0)=h0'Y\dot{x}(0);\ \dot{y}(0)=h0'Y\dot{y}(0)$

The whirl stability can be verified with the orbit method [18,32], that considers the nonlinearities of the problem.

To analyze the dynamic properties of the thrust bearing the Reynolds equation (8) is solved in time together with the axial spindle equation of motion (18):

$$mr\ddot{z}+Ft=F$$

in which F_t is the reaction force of the thrust bearing, F is the external axial force applied to the shaft and m_r is the rotor mass. The reaction force is calculated by integrating the pressure over the thrust surface.

CONCLUSIONS

This paper investigated four prototypes of high speed spindles and rotors supported by gas bearings. The radial and axial bearings were designed using a numerical program that simulates the pressure distribution inside the air clearance both in static and in dynamic conditions.

Different were the design priorities: stiffness and load capacity for the pneumatic spindle and the electro-spindle and stability for the textile spindle and the mesoscopic spindle. Two different methods were employed to increase stability: the introduction of external damping in the textile spindle and the modification of the film geometry in the mesoscopic spindle.

The experimental activity carried out demonstrated the suitability of gas bearings for different applications and made it possible to verify the numerical models. Experimental testing played an essential role in identifying the numerical models and in the same time the models, once identified, made it possible to save time in the design process. In particular the supply holes configuration of the prototypes was optimized using the numerical models to choose their number, diameter and disposition. For stability optimization of the mesoscopic spindle the models were used as virtual test benches to compare different geometries of bushing with non-circular profile. The results of the tests on rubber rings were used in numerical models to investigate their effect on stability.

The prototypes developed operated in stable conditions in the speed range expected. Future investigations will verify the stability at higher speeds.

NOMENCLATURE

b : ratio of critical pressure to admission pressure, b=0.528

c_d : supply hole discharge coefficient

cOR : damping coefficient of the O-ring

c_s : supply hole conductance

D : journal bearing diameter

d_s : supply hole diameter

F_t : external axial force on thrust bearing

Fx,Fy : external forces on rotor

G : air mass flow rate through the supply hole

h : local air clearance

$h0$: clearance with rotor in centred position

J_G : transverse rotor moment of inertia, calculated respect to the centre of mass

J_P : polar moment of inertia of rotor

k_T : temperature coefficient, k_T=293/$T0$------√

kOR : stiffness coefficient of the O-ring

L : bearing axial length

mb : bushing mass

mr : rotor mass

n,m : number of nodes along axial and circumferential directions

p : pressure

p_a : ambient pressure

p_s : bearing supply pressure

$q$inlet : mass flow rate per unit surface

R : journal bearing radius

r,θ,z : cylindrical coordinates

R^0 : gas constant, in calculations R^0=287.6 m^2/s^2K

Re : Reynolds number calculated at supply port section, $Re=4G/(\pi d_s \mu)$

Re^* : modified Reynolds number, $Re^*=\rho\omega h_0^2/\mu$

S : supply hole cross section

T^0 : absolute temperature, in calculations T^0=288 K

u,v : mean velocity components in z- and -direction

x,y,z : cartesian coordinates

zG : center of mass axial coordinate

zF : axial coordinate of the external force on the rotor

ex, ey : rotor eccentricities

Δt : time step

Δz : mesh size in axial direction

$\Delta\theta$: mesh size in circumferential direction

Λ : bearing number, $\Lambda=6\mu\omega/p_a\cdot(D/2h_0)^2$

χ : dynamic rotor unbalance

ε : static rotor unbalance

ε_x, ε_y : rotor eccentricity ratios

γ : whirl ratio, $\gamma=\upsilon/\omega$

ϕ : angle between static and dynamic unbalance

μ : dynamic viscosity, in calculations $\mu=17.89\cdot10^{-6}$ Pa·s

υ : whirl frequency

ρ_N : air density in normal conditions

ω : rotor angular speed

REFERENCES

1. D. D. Fuller, 1984Theory and practice of lubrication for engineers, John Wiley and Sons, New York.
2. T. Waumans, J. Peirs, F. Al-Bender, D. Reynaerts, 2011Aerodynamic bearing with a flexible, damped support operating at 7.2 million DN, J. Micromech. Microeng. 21104014 EOF
3. D. A. Boffey, D. M. Desay, 1980An experimental investigation into the

rubber-stabilization of an externally-pressurized air-lubricated thrust bearing, ASME Trans. Journal of lubrication technology, 1026570

4. K. Czolczynski, 1994Stability of flexibly mounted self acting gas journal bearings, Nonlinear Science, Part B, Chaos and nonlinear Mechanics, 7286299

5. R. Zhang, H. S. Chang, 1995A new type of hydrostatic/hydrodynamic gas journal bearing and its optimization for maximum stability, STLE Tribology Transactions, 383589594

6. D. Yang-W, C. Chen-H, Y. Kang, R. Hwang-M, S. Shyr-S, 2009Influence of orifices on stability of rotor-aerostatic bearing system, Tribology International, 4212061219

7. Westwind2008aOnline available: http://www.westwind-airbearings.com/specialist/wafer Grinding.html

8. Westwind2008bOnline available: http://www.westwind-airbearings.com/pcb/overview.html

9. APT, Air Bearing Precision Technology.Catholic University of Leuven, Belgium. Online available: http://www.mech.kuleuven.be/industry/spin/APT/default_en.

10. S. Ohishi, Y. Matsuzaki, 2002Experimental investigation of air spindle unit thermal characteristics. Precision Engineering, 264957

11. Moore Precision Tools2001Nanotechnology Systems.

12. Precitech Precision,2001Nanoform® 350 Technical Overview and Unsurpassed Part Cutting Results.

13. Toshiba Machine Co. Ltd.2002High Precision Aspheric Surface Grinder.

14. G. Belforte, F. Colombo, T. Raparelli, A. Trivella, V. Viktorov, 2008aHigh speed electrospindle running on air bearings: design and experimental verification, Meccanica, 43591600

15. G. Belforte, F. Colombo, T. Raparelli, V. Viktorov, A. Trivella, 2006An experimental study of high speed rotor supported by air bearings: test rig and first experimental results, Tribology International, 39839845

16. L. Della Pietra, G. Adiletta, 2002The squeeze film damper over four decades of investigations: part 1. Characteristics and operating features, Shock Vib. Dig., 34326

17. G. Belforte, F. Colombo, T. Raparelli, A. Trivella, V. Viktorov, 2008bHigh speed rotors with air bearings mounted on flexible supports: test bench and experimental results. ASME Journal of Tribology, 130

18. G. Belforte, T. Raparelli, V. Viktorov, 1999Theoretical investigation of fluid inertia effects and stability of self-acting gas journal bearings. ASME

Journal of Tribology, 121836 EOF843 EOF

19. G. Belforte, T. Raparelli, V. Viktorov, A. Trivella, 2007Discharge coefficients of orifice-type restrictor for aerostatic bearings, Tribology International, 40512521
20. P. C. Mishra, R. K. Pandley, K. Athre, 2007Temperature profile of an elliptic bore journal bearing, Tribology International, 40453458
21. H. Hashimoto, 1992Dynamic characteristic analysis of short elliptical journal bearings in turbulent inertial flow regime, STLE Tribology Transactions, 354619626
22. H. Hashimoto, K. Matsumoto, 2001Improvement of operating characteristics of high speed hydrodynamic journal bearings by optimum design: Part I- formulation of methodology and its application to elliptical bearing design, ASME Journal of Tribology, 123305312
23. N. Z. Wang, C. L. Ho, K. C. Cha, 2000Engineering optimum design of fluid film lubricated bearings, Journal of Tribology Transactions, 433377386
24. L. J. Read, R. D. Flack, 1987Temperature, pressure and film thickness measurements for an offset half bearing. Wear; 1172197210
25. N. M. Ene, F. Dimofte, Keith Jr, T. G. , 2008aA dynamic analysis of hydrodynamic wave journal bearings, STLE Tribology Transactions, 5118291
26. N. M. Ene, F. Dimofte, Keith Jr, T. G. , 2008bA stability analysis for a hydrodynamic three-wave journal bearing, Tribology International, 415434442
27. R. Sehgal, K. N. S. Swamy, K. Athre, S. Biswas, 2000A comparative study of the thermal behaviour of circular and non-circular journal bearings. LubSci, 124329344
28. F. Dimofte, 1995aWave journal bearing with compressible lubricant- Part I : The wave bearing concept and a comparison to the plain circular bearing, Tribology Transactions, 381153160
29. F. Dimofte, 1995bWave journal bearing with compressible lubricant- Part II : A comparison of the wave bearing with a groove bearing and a lobe bearing, Tribology Transactions, 382364372
30. V. Viktorov, G. Belforte, T. Raparelli, F. Colombo, 2009Design of non-circular gas bearings for ultra-high speed spindle. World Tribology Congress, Kyoto, 611Sept. C1-212.
31. V. Castelli, J. Pirviks, 1968Review of numerical methods in gas bearing film analysis. Journal of Lubrication Technology, 777792
32. F. Colombo, T. Raparelli, V. Viktorov, 2009Externally pressurized gas bearings: a comparison between two supply holes configurations. Tribology International, 42

Chapter 4

REVIEW OF THE NEW COMBUSTION TECHNOLOGIES IN MODERN GAS TURBINES

M. Khosravy el-Hossaini[1]

[1]Research Institute of Petroleum Industry, Iran

INTRODUCTION

The combustion chamber is the most critical part of a gas turbine. The chamber had to be designed so that the combustion process to sustain itself in a continuous manner and the temperature of the products is sufficiently below the maximum working temperature in the turbine. In the conventional industrial gas turbine combustion systems, the combustion chamber can be divided into two areas: the primary zone and the secondary zone. The primary zone is where the majority of the fuel combustion takes place. The fuel must be mixed with the correct amount of air so that a stoichiometric mixture is present. In the secondary zone, unburned air is mixed with the combustion products to cool the mixture before it enters the turbine. In some design, there is an intermediate zone where help secondary zone to eliminate the dissociation products and burn-out soot.

The majority of the combustors are developed base on diffusion flames as they are very stable and fuel flexibility option. In a diffusion flame, there will be always stoichiometric regions regardless of overall stoichiometry. The main disadvantage of diffusion-type combustor is the emission as high temperature of the primary zone produced larger than 70 ppm NOx in burning natural gas and more than 100 ppm for liquid fuel [1]. Several techniques have been tried in order to reduce the amount of NOx produced in conventional combustors. In general, it is difficult to reduce NOx emissions while maintaining a high combustion efficiency as there is a tradeoff between NOx production and CO/UHC production.

In some recent installations, the premixed type of combustion has been selected to reduce NOx emissions bellow 10 ppm. Apart from the flame type change, there are some method such as "wet diffusion combustion", FGR[1] - and SCR[2] - . In an example of wet combustion, a nuzzle through which steam is injected is provided in the vicinity of the fuel injector. The level of NOx emission is controlled by the amount of steam. However, there is a limit on the increasing the steam flow rate as cause corresponding considerable CO emission. Further more, preparing pure steam in the required injection condition increases operational costs. Now a days, wet combustion rarely applies due to water consumption and the penalty of reduced efficiency. Post Combustion treatments such as SCR are those which convert NOx compounds to nitrogen or absorb them from flue gas. These methods are relatively in expensive to install but does not achieve NOx removal levels better than modern gas turbine combustor.

In this chapter, a short introduction of combustion process and then a description of some new pioneer combustor have been presented. As gas turbine manufacturers are looking for continuous operation or stable combustion, satisfactory emission level, minimum pressure loss and durability or life. Hence, the advanced combustor might include all of these criteria, so some of them are selected to discuss in details.

THE COMBUSTION PROCESS

Type of Combustion Chamber

The diffusion and premixed flame are two main type of combustion, which are using in gas turbines. Apart from type of flame, there are two kind of combustor design, annular and tubular. The annular type mostly recommended in the propulsion of aircraft when small cross section and low weight are important parameters. Can or tubular combustors are cheaper and several of them can be adjusted for an industrial engine identically. Although there are different types of combustors, but generally, all combustion chambers have a diffuser, a casing, a liner, a fuel injector and a cooling arrangement. An entire common layout is visualized in figure 1.

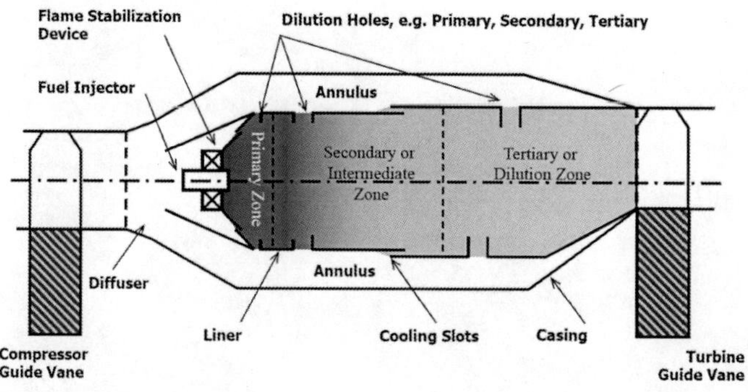

Figure 1. The layout of the combustion chamber.

Flame Stabilization

After the fuel has been injected into the air flow, the flow will enter the flame region. It does this with quite a high velocity, so to make sure the flame isn't blown away; suitable flame stabilization techniques must be applied. First, the high velocity of flow will be responsible for a pressure drop[3] - . Secondly, the flame in the combustion chamber cannot survive if the air has a high velocity. So combustion chambers benefit from diffusers to slow down the air flow. There are two normal kinds of flame stabilizers: bluff-body flame holders and swirlers.

The shape of the bluff–body flame holder affects the flow stability characteristics through the influence on the size and shape of the wake region. Since the flame stabilization depends on size of the zone of recirculation behind the bluff–body, different geometries such as triangular, rectangular, circular and more complex shapes are being use. One of the basic problem of bluff-body flame holders is a considerable effect on pressure loss. Figure 2 shows a high speed image of three flame holders in atmospheric condition.

Figure 2. High speed images of the circular cylinder (top), square cylinder (middle) and V gutter (bottom) at Re = 30,000 and stoichiometric mixture [2].

Flow reversal can be applied in the primary zone. The best way to reverse the flow is to swirl it through using swirlers. The two most important types of swirlers are axial and radial. The advantage of flow reversal is that the flow speed varies a lot. So there will be a point at which the airflow velocity matches the flame speed where a flame could be stabilized. The degree of swirl in the flow is quantified by the dimensionless parameter, *Sn* known as the swirl number which is defined as:

$$Sn = \frac{G_\theta}{G_x r}$$

Where:

$$G_\theta = \int_0^\infty \left(uw + \overline{u'w'}\right) r^2 dr \qquad G_x = \int_0^\infty \left(u^2 + \overline{u'^2} + (p - p_\infty)\right) r dr$$

As this equation requires velocity and pressure profile of fluid, researchers proposed various expressions for calculating the swirl number. Indeed, the swirl number is a non-dimensional number representing the ratio of axial flux of angular momentum to the axial flux of axial momentum times the equivalent nozzle radius [3]. Tangential entry, guided vanes and direct rotation are three principal methods for generating swirl flow.

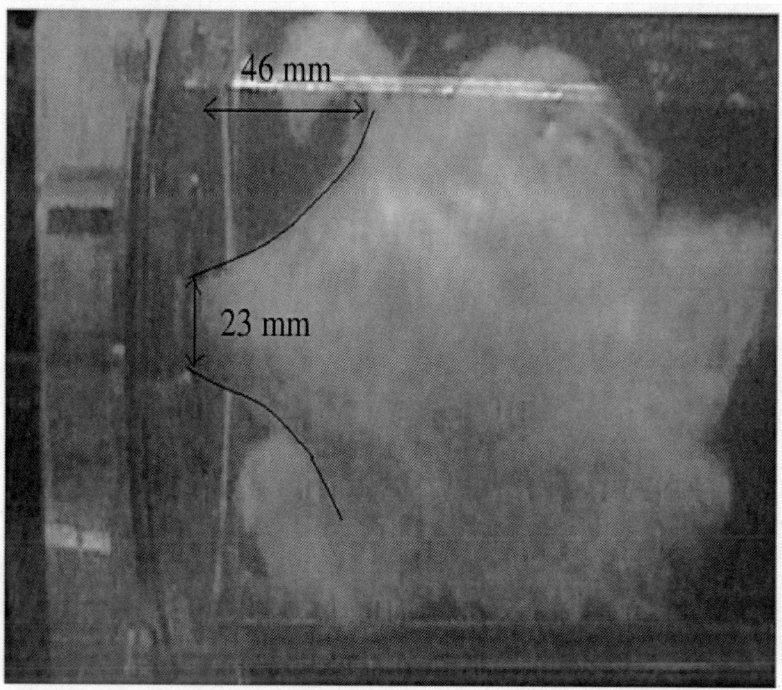

Figure 3. Photo of 60° flat guided vane swirler [4].

Type of Flame

Most of the literatures divide combustible mixture into three categories as premixed, non-premixed and partially premixed combustion. If fuel and oxidizer are mixed prior ignition, then premixed flame will propagate into the unburned reactants. If fuel and air mix at the same time and same place as they react, the diffusion or non-premixed combustion will appear. Partially premixed combustion systems are premixed flames with non-uniform fuel-oxidizer mixtures.

Gas turbines' manufacturers traditionally tend to use diffusion flame where fuel mixes with air by turbulent diffusion and the flame front stabilized in the locus of the stoichiometric mixture. The temperature of reactant is as high as 2000 °C, so the acceptable temperature at the combustor walls and turbine blades would be provide by diluted air. Although the non-premixed mixture in gas turbine combustors shows more stability in operation than premixed mixtures, but their shortcoming is high level of nitrogen oxide emission. Two most common ways of emission reduction are water injection and catalytic converter. However, the former technique is not capable of reducing NOx to the expected level at many sites, while SCR adds complexity and expense to any project.

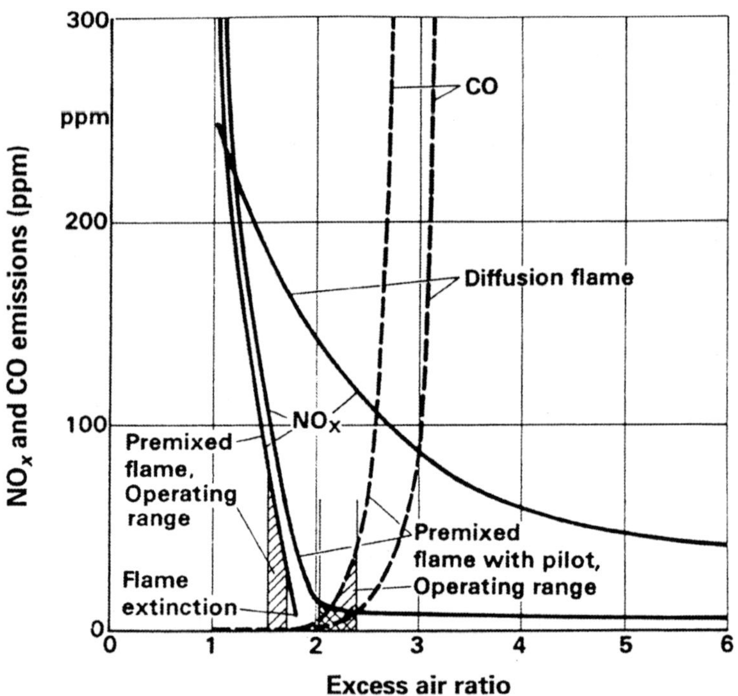

Figure 4. Operating range of premixed flames [5].

The idea of Dry Low NOx (DLN) systems proposed base on lean premixed combustion to reduce flame temperature by a non-stoichiometric mixture. Premixed systems can be operated at a much lower equivalence ratio such that the flame temperature and thermal NOx production throughout the system are decreased comparing with a diffusion system. The disadvantage of premixed systems is flame stability, especially at low equivalence ratios. Also, there is a tendency for the flame to flashback. Indeed, the current challenge of GT's developers is proposing a fuel flexible combustor for a stable combustion in all engine loads. The narrow range of fuel/air mixtures between the production of excessive NOx and excessive CO is illustrated in figure 4. NOx reduces by lowering flame temperature in a leaner mixture but CO, and unburned hydrocarbons (UHC) would increase contradictorily.

By increasing combustion residence time (volume) and preventing local quenching, CO and UHC will dissociate to CO_2 and the other products. CO burns away more slowly than the other radicals, so to obtain very low level emission such as 10 ppm; it requires over 4 ms. As shown in figure 5, below 1100 °C the CO reaction becomes too slow to effectively remove the CO in an improved combustion chamber. The residence time usually does not change much on part-load because the normalized flow approximately remains constant with a variable loading.

$$NF = \dot{m}TP$$

Where $\dot{m}$ is the mass flow, T is combustion bulk temperature and P is combustor pressure. This will set a lower limit for the length of the primary zone in a DLN combustion system.

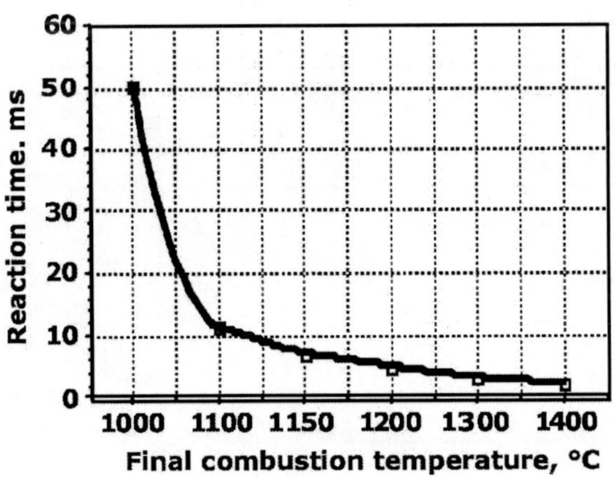

Figure 5. Calculated reaction time to achieve a CO concentration of 10 ppm in a commercial gas turbine exhaust [6].

FUEL

One of the features of heavy-duty gas turbines is a wide fuel capability. They can operate with vast series of commercial and process by-product fuels such as natural gas, petroleum distillates, gasified coal or biomass, gas condensates, alcohols, ash-forming fuels. In a review article, Molière offered essential aspects of fuel/ machine interactions in thermodynamic performance, combustion and gaseous emission [7]. To sequester and store the CO_2 of fossil fuel, some new research projects aim to assess the combustion performances of alternative fuels for clean and efficient energy production by gas turbines. Another objective is to extend the capability of dry low emission gas turbine technologies to low heat value fuels produced by gasification of biomass and H_2 enriched fuels [8-10]. Significant quantity of hydrogen in fuel has the benefit of high calorific value, but the disadvantage of high flame speed and very fast chemical times. To classify gas turbine's fuels, a common way is to split them between gas and liquid fuels, and within the gaseous fuels, to split by their calorific value as shown in table 1.

Table 1. Classification of fuels [11].

	Typical composition	Lower Heating Value kJ/N^{m}3	Typical specific fuels
Ultra/Low LHV gaseous fuels	$H_2 < 10\%$	< 11,200 (< 300)	Blast furnace gas (BFG), Air blown IGCC, Biomass gasification
	$CH_4 < 10\%$		
	N_2+CO > 40%		
High hydrogen gaseous fuels	$H_2 > 50\%$	5,500-11,200 (150-300)	Refinery gas, Petrochemical gas, Hydrogen power
	$C_xH_y = 0\text{-}40\%$		
Medium LHV gaseous fuels	$CH_4 < 60\%$	11,200-30,000	Weak natural gas, Landfill gas, Coke oven gas, Corex gas
	$N_2+CO_2 = 30\text{-}50\%$		
	$H_2 = 10\text{-}50\%$		
Natural gas	$CH_4 = 90\%$	30,000-45,000	Natural gas Liquefied natural gas
	$C_xH_y = 5\%$		
	Inert = 5%		

High LHV gaseous fuels	CH_4 and higher hydrocarbons	45,000-190,000	Liquid petroleum gas (butane, propane) Refinery off-gas
	$C_xH_y > 10\%$		
Liquid fuels	C_xH_y, with $x > 6$	32,000-45,000	Diesel oil, Naphtha Crude oils, Residual oils, Bio-liquids

NEW COMBUSTION SYSTEMS FOR GAS TURBINES

Next-generation gas turbines will operate at higher pressure ratios and hotter turbine inlet temperatures conditions that will tend to increase nitrogen oxide emissions. To conform to future air quality requirements, lower-emitting combustion technology will be required. In this section, a number of new combustion systems have been introduced where some of them could be found in the market, and the others are under development.

Trapped Vortex Combustion (TVC)

The trapped vortex combustor (TVC) may be considered as a promising technology for both pollutant emissions and pressure drop reduction. TVC is based on mixing hot combustion products and reactants at a high rate by a cavity stabilization concept. The trapped vortex combustion concept has been under investigation since the early 1990's. The earlier studies of TVC have been concentrated on liquid fuel applications for aircraft combustors [12].

The trapped vortex technology offers several advantages as gas turbines burner:

- It is possible to burn a variety of fuels with medium and low calorific value.
- It is possible to operate at high excess air premixed regime, given the ability to support high-speed injections, which avoids flashback.
- NOx emissions reach extremely low levels without dilution or post-combustion treatments.
- Produces the extension of the flammability limits and improves flame stability.

Flame stability is achieved through the use of recirculation zones to provide a continuous ignition source which facilitates the mixing of hot combustion products with the incoming fuel and air mixture [13]. Turbulence occurring in a TVC combustion chamber is "trapped" within a cavity where reactants are injected and efficiently mixed. Since part of the combustion occurs within the recirculation zone, a "typically" flameless regime can be achieved, while a trapped turbulent vortex may provide significant pressure drop reduction [14]. Besides this, TVC is having the capability of operating as a staged combustor if the fuel is injected into both the cavities and the main airflow. Generally, staged combustion systems are having the potential of achieving about 10 to 40% reduction in NOx emissions [15]. It can also be operated as a rich-burn, quick-quench lean-burn (RQL) combustor when all of the fuel is injected into the cavities [16].

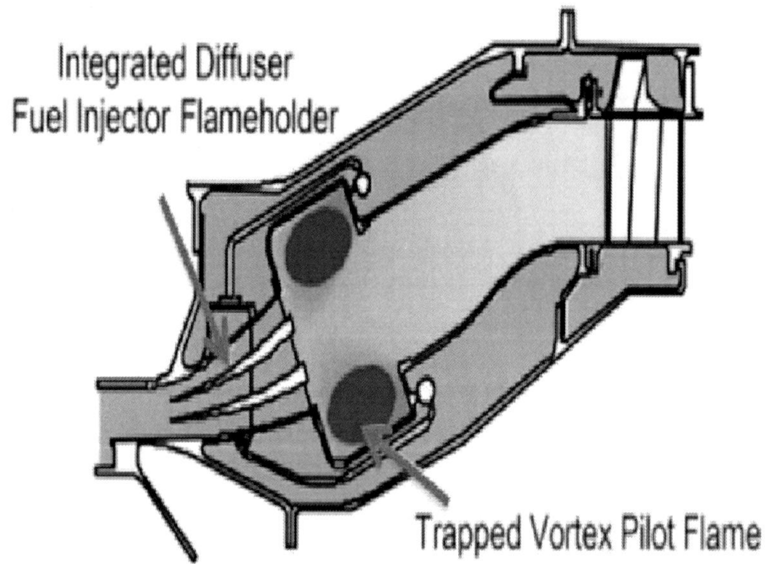

Figure 6. Trapped vortex combustor schematic.

An experiment in NASA with water injected TVC demonstrated a reduction in NOx by a factor three in a natural gas fueled and up to two in a liquid JP-8 fueled over a range in water/fuel and fuel/air ratios [17]. Replacement of natural gas fuel with syngas and hydrogen fuels has been studied numerically by Ghenai et al. [18]. The effects of secondary air jet momentum on cavity flow structure of TVC have been studied recently by Kumar and Mishra [19]. Although the actual stabilization mechanism facilitated by the TVC is relatively simple, a number of experiments and numerical simulations have been performed to enhance the stability of reacting flow inside trapped vortex. Xing et al. experimentally investigated lean blow-out of several combustors and the performance of slight temperature-raise in a single trapped vortex [20, 21]. In an experimental laboratory research, Bucher et al. proposed a new design for lean-premixed trapped vortex combustor [22].

Rich Burn, Quick- Mix, Lean Burn (RQL)

Lean direct injection (LDI) and rich-burn/quick-quench/lean-burn (RQL) are two of the prominent low-emissions concepts for gas turbines. LDI operates the primary combustion region lean, hence, adequate flame stabilization has to be ensured; RQL is rich in the primary zone with a transition to lean combustion by rapid mixing with secondary air downstream. Hence, both concepts avoid stoichiometric combustion as much as possible, but flame stabilization and combustion in the main heat release region are entirely different. Relative to aviation engines, the need for reliability and safety has led to a focus on LDI of liquid fuels [23]. However, RQL combustor technology is of growing interest for stationary gas turbines due to the attributes of more effectively processing of fuels with complex composition. The concept of RQL was proposed in 1980 as a significant effort for reducing NOx emission [24].

It is known that the primary zone of a gas turbine combustor operates most effectively with rich mixture ratios so, a "rich-burn" condition in the primary zone enhances the stability of the combustion reaction by producing and sustaining a high concentration of energetic hydrogen and hydrocarbon radical species. Secondly, rich burn conditions minimize the production of nitrogen oxides due to the relative low temperatures and low population of oxygen containing intermediate species. Critical factors of a RQL that need to be considered are careful tailoring of rich and lean equivalence ratios and very fast cooling rates. So the combustion regime shifts rapidly from rich to lean without going through the high NOx route as shown in figure 7. The drawback of this technology is increased hardware and complexity of the system.

The mixing of the injected air takes the reaction to the lean-burn zone and rapidly reduces their temperature as well. On the other hand, the temperature must be high enough to burn CO and UHC. Thus, the equivalence ratio for the lean-burn zone must be carefully selected to satisfy all emissions requirements. Typically the equivalence ratio of fuel-rich primary zone is 1.2 to 1.6 and lean-burn combustion occurs between 0.5 and 0.7 [25].

Turbulent jet in a cross-flow is an important characteristic of RQL; so many researches have been conducted to improve it. The mixing limitation in a design of RQL/TVC combustion system addressed by Straub et al. [26]. Coaxial swirling air discussed experimentally by Cozzi and Coghe [27]. Furthermore, an experimental study of the effects of elevated pressure and temperature on jet mixing and emissions in an RQL reported by Jermakian et al. [28]. Fuel flexible combustion with RQL system is an interest of turbine manufacturer. GE reported results of a RQL test stand in their integrated gasification combined cycle (IGCC) power plants program [29, 30]. The test of Siemens-Westinghouse Multi-Annular Swirl Burner (MASB) was successfully performed at the University of Tennessee Space Institute in Tullahoma [31]. Others, such as references [32-35] utilize CFD to investigate the performance of RQL combustor.

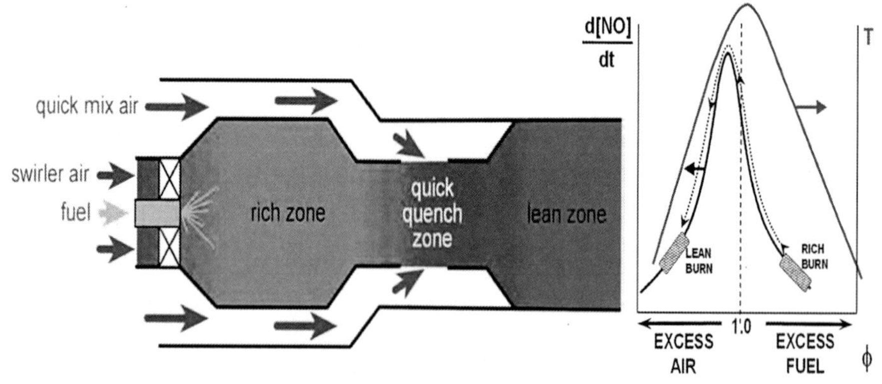

Figure 7. Rich-Burn, Quick-Mix, Lean-Burn combustor.

Staged Air Combustion

The COSTAIR[4] - combustion concept uses continuously staged air and internal recirculation within the combustion chamber to obtain a stable combustion with low NOX and CO emissions. Research work on staged combustors started in the early 1970s under of the Energy Efficient Engine (E^3) Program in the USA [36] and now widely used in industrial engines burning gaseous fuels, in both axial and radial configurations. The aero-derived GE LM6000 and CFM56-5B as well as RR211 DLE industrial engine employ staged combustion of premixed gaseous fuel/air mixtures. Recently, a research project proposed a COSTAIR burner system optimized for low calorific gases within a micro gas turbine [37].

The principle of staged air combustion is illustrated in Figure 8. It consists of a coaxial tube; the combustion air flows through the inner tube and the fuel through the outer cylinder ring. The combustion air is continually distributed throughout the combustion chamber by an air distributor with numerous openings on its contour, and fuel enters by several jets arranged around the air distributor.

The COSTAIR burner has the advantages of operating in full diffusion mode or in partially premixed mode. The heat is released more uniformly throughout the combustion chamber also the recirculated gas absorb some of the heat of combustion. It capable to work stable at cold combustor walls as well as high air ratio. Experimental measurements show that this combustion system allows clean exhaust. For instance, in an experimental research project of European Commission [39], NOx emission values was in the range of 2-4 ppm at an air ratio of 2.5 over different loading. Furthermore, the corresponding CO emission was less than 7 ppm.

Staged combustion can occur in either a radial or axial pattern, but in either case the goal is to design each stage to optimize particular performance aspects. The main advantages or major drawbacks of each type have been discussed by Lefebvre [25].

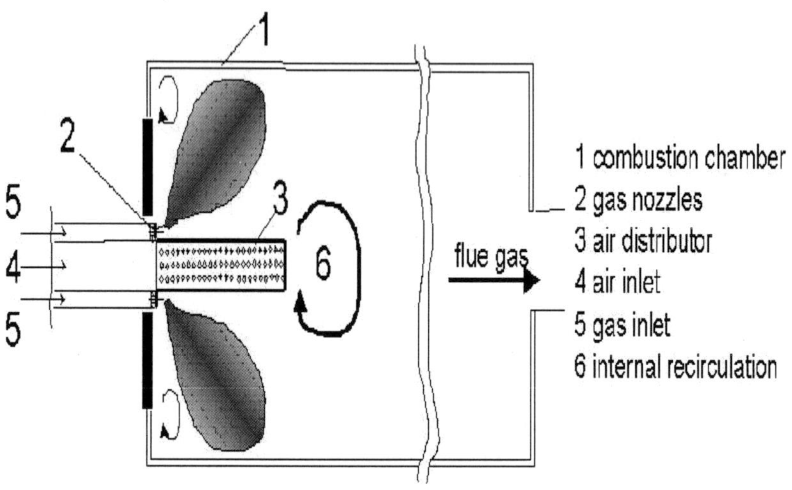

Figure 8.COSTAIR combustion concept [38].

Mild Combustion

Heat recirculating combustion was clearly described by Weinberg as a concept for improving the thermal efficiency [40]. In 1989, a surprising phenomenon was observed during experiments with a self-recuperative burner. At furnace temperatures of 1000°C and about 650°C air preheated temperature; no flame could be seen, but the fuel was completely burnt. Furthermore, the CO and NOx emissions from the furnace were considerably low [41]. Different combustion zones against rate of dilution and oxygen content is shown in figure 9. In flameless combustion, the oxidation of fuel occurs with a very limited oxygen supply at a very high temperature. Spontaneous ignition occurs and progresses with no visible or audible signs of the flames usually associated with burning. The chemical reaction zone is quite diffuse, and this leads to almost uniform heat release and a smooth temperature profile. All these factors could result in a much more efficient process as well as reducing emissions.

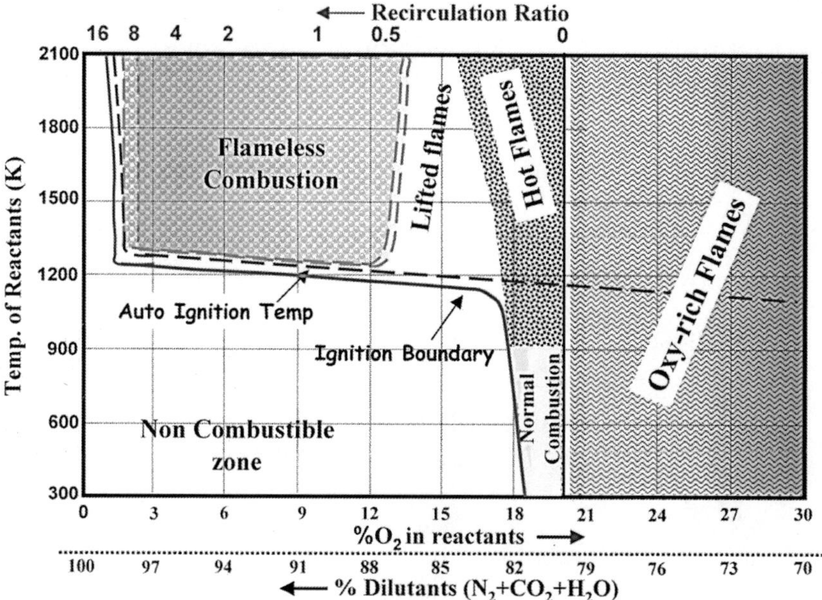

Figure 9. Different combustion regimes [64].

Flameless combustion is defined where the reactants exceed self-ignition temperature as well as entrain enough inert combustion products to reduce the final reaction temperature [42]. In the other word, the essence of this technology is that fuel is oxidized in an environment that contains a substantial amount of inert (flue) gases and some, typically not more than 3–5%, oxygen. Several different expressions are used to identify similar though such as HiTAC[5] - , HiCOT[6] - , MILD[7] - combustion, FLOX[8] - and CDC[9] - . HiTAC refers to increase the air temperature by preheating systems such as regenerators. HiCOT commonly belongs to the wider sense, which exploits high-temperature reactants; therefore, it is not limited to air. A combustion process is named FLOX or MILD when the inlet temperature of the main reactant flow is higher than mixture autoignition temperature and the maximum allowable temperature increase during combustion is lower than mixture autoignition temperature, due to dilution [42]. The common key feature to achieve reactions in CDC mode (non-premixed conditions) is the separation and controlled mixing of higher momentum air jet and the lower momentum fuel jet, large amount of gas recirculation and higher turbulent mixing rates to achieve spontaneous ignition of the fuel to provide distributed combustion reactions [43].Figure 10 schematically shows a comparison between conventional burner and flameless combustion.

To recap, the main characteristics of flameless oxidation combustion are:

- Recirculation of combustion products at high temperature (normally > 1000 °C),
- Reduced oxygen concentration at the reactance,
- Low Damköhler number (Da[10] -),
- Low stable adiabatic flame temperature,
- Reduce temperature peaks,
- Highly transparent flame,
- Low acoustic oscillation and
- Low NOx and CO emissions.

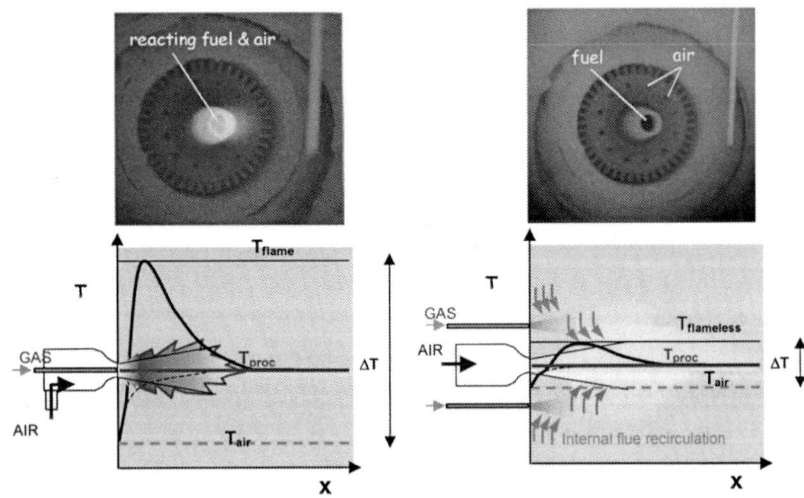

Figure 10. Flame (left) and flameless (right) firing.

In spite of a number of activities for industrial furnaces, the application of flameless combustion in the gas-turbine combustion system is in the preliminary phase [44]. The results from techno-economic analysis of Wang et al. showed that the COSTAIR and

FLOX cases had technical and economic advantages over SCR [45]. Luckerath, R., et al., investigated flameless combustion in forward flow configuration in elevated pressure up to 20atm for application to gas turbine combustors [44, 46]. In a novel design of Levy et al. that named FLOXCOM, flameless concept has been proposed for gas turbines by establishing large recirculation zone in the combustion chamber [47, 48]. Lammel et al. developed a FLOX combustion at high power density and achieved low NOx and CO levels [49]. The concept of colorless distributed combustion has been demonstrated by Gupta et al. for gas turbine application in a number of publications [43, 50-55].

Surface Stabilized Combustion

One specification of gas turbine combustor is higher thermal intensity range (at least 5 MW/m^3-atm) than industrial furnaces which operate at thermal intensity of less than 1 MW/m^3-atm. Therefore, designs of gas turbine's combustors are based on turbulent flow concept, except a technology named NanoSTAR from Alzeta Corporation. Alzeta reported the proof-of-concept of high thermal intensity laminar surface stabilized flame by using a porous metal-fiber mat since 2001 [56-58]. Lean premixed combustion technology is limited by the apparition of combustion instabilities, which induce high pressure fluctuations, which can produce turbine damage, flame extinction, and CO emissions [59]. However, full scale test of NanoSTAR demonstrated low emissions performance, robust ignition and extended turndown ratio [60]. In particular, the following characteristics form the key specifications of NanoSTAR for distributed power generation gas turbine combustors [61]:

- The combustor fuel is limited to natural gas.
- Total combustor pressure drop limited to 2-4% of the system pressure.
- Operation at combustion air preheat temperatures up to 1150°F.
- Volumetric firing rates approaching 2 $MMBtu/hr/atm/ft^3$.
- Turbine Rotor Inlet Temperatures (TRIT) over 2200°F (valid for the Mercury 50, although Allison has operated combustors at 2600°F).
- Operation with axial combustors or external can combustors.

- Expected component lifetimes of 30,000 hours for industrial turbines.
- A single prototype burner Porous burner which sized to fit inside an annular combustion liner (about 2.5 inches in diameter by 7 inches in length) is shown in figure 11 with its arrangement in a typical combustor.

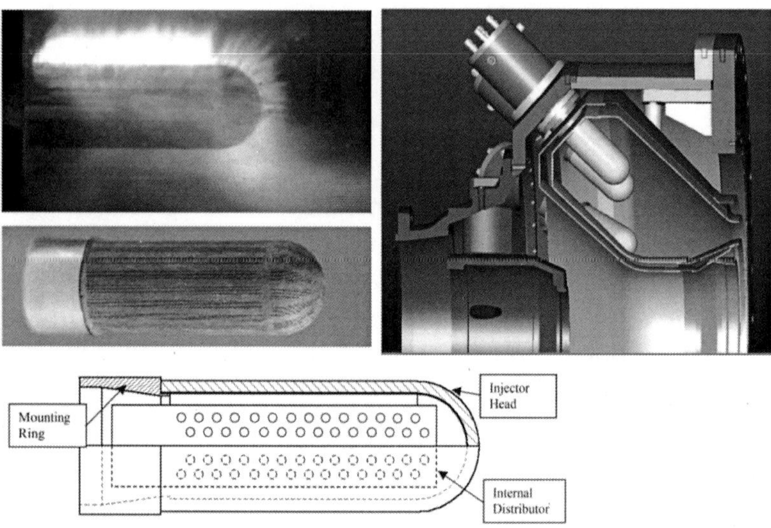

Figure 11. NanoSTAR burner and its arrangement in a canted combustion system [62].

The operation of this type of surface stabilized combustion is characterized by the schematic inFigure 12(left), which shows premixed fuel and air passing through the metal fiber mat in two distinct zones. Premixed fuel comes through the low conductivity porous and burns in narrow zones, A, as it leaves the surface. Under lean conditions this will manifest as very short laminar flamelets, but under rich conditions the surface combustion will become a diffusion dominated reaction stabilized just over a millimeter above the metal matrix, which proceeds without visible flame and heats the outer surface of the mat to incandescence. Secondly, adjacent to these radiant zones, the porous plate is perforated to allow a high flow of the premixed fuel and air. This flow forms a high intensity flame, B, stabilized by the radiant zones so, it is possible to achieve very high fluxes of energy, up to 2MMBtu/hr/ft^2 [63]. A picture of an atmospheric burner in operation clearly shows the technology in action (right offigure 12).

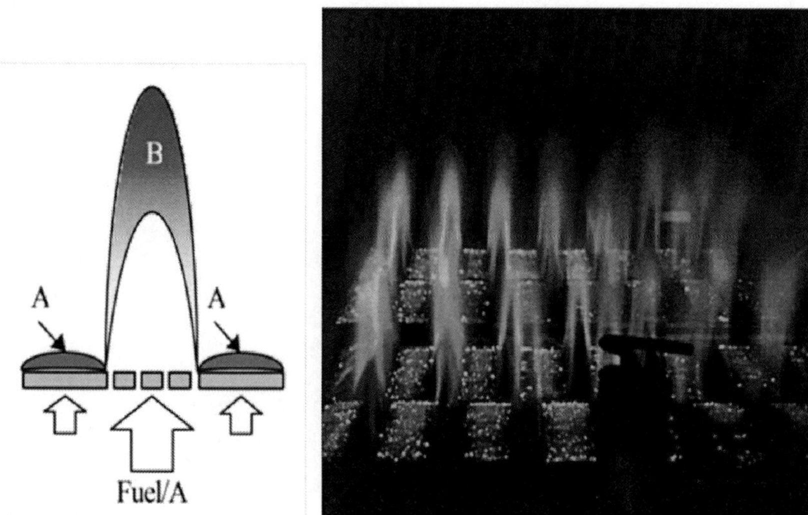

Figure 12. Surface stabilized burner pad firing at atmospheric conditions.

The specific perforation arrangement and pattern control the size and shape of the laminar flamelets. The perforated zones operate at flow velocities of up to 10 times the laminar flame speed producing a factor of ten stretch of the flame surface and resulting in a large laminar flamelets. The alternating arrangement of laminar blue flames and surface combustion, allows high firing rates to be achieved before flame liftoff occurs, with the surface combustion stabilizing the long laminar flames by providing a pool of hot combustion radicals at the flame edges.

CONCLUSION

A review of technologies for reducing NOx emissions as well as increasing thermal efficiency and improving combustion stability has been reported here. Trade-offs when installing low NOx burners in gas turbines include the potential for decreased flame stability, reduced operating range and more strict fuel quality specifications. In the other word, although, the turbine inlet temperature is the major factor determining the overall efficiency of the gas turbine but higher inlet temperatures will result in larger NOx emissions. So the essential requirement of new combustor design is a trade-off between low NOx and improved efficiency.

Notes

1. Flue Gas Recirculation
2. Selective Catalytic Reduction
3. This pressure drop is named the cold loss.
4. COntinuousSTaged Air
5. High Temperature Air Combustion
6. High-temperature Combustion Technology
7. Moderate or Intense Low-oxygen Dilution
8. FLamelessOXidation
9. Colorless Distributed Combustion
10. A dimensionless number, equal to the ratio of the turbulence time scale to the time it takes chemical reaction.

REFERENCES

1. Chambers A and Trottier S (2007) Technologies for Reducing Nox Emissions from Gas-Fired Stationary Combustion Sources. Alberta Research Council, Edmonton, Canada
2. Kiel B, Garwick LK, Gord JR, Miller J, Lynch A, Hill R and Phillips S (2007) A Detailed Investigation of Bluff Body Stabilized Flames. 45th AIAA Aerospace Sciences Meeting and Exhibit.
3. Gupta AK, Lilley DG and Syred N (1984) Swirl Flows, Abacus Press.
4. Jaafar MNM, Jusoff K, Osman MS and Ishak MSA (2011) Combustor Aerodynamic Using Radial Swirler.International Journal of the Physical Sciences.6: 3091 - 3098.
5. Moore MJ (1997) Nox Emission Control in Gas Turbines for Combined Cycle Gas Turbine Plant. Proceedings of the Institution of Mechanical Engineers, Part A: Journal of Power and Energy. 211: 43-52.

6. Kajita S and DallaBetta R (2003) Achieving Ultra Low Emissions in a Commercial 1.4 Mw Gas Turbine Utilizing Catalytic Combustion. Catalysis Today. 83: 279-288.
7. Molière M (2000) Stationary Gas Turbines and Primary Energies: A Review of Fuel Influence on Energy and Combustion Performances. International Journal of Thermal Sciences. 39: 141-172.
8. Gupta KK, Rehman A and Sarviya RM (2010) Bio-Fuels for the Gas Turbine: A Review. Renewable and Sustainable Energy Reviews. 14: 2946-2955.
9. Gökalp I and Lebas E (2004) Alternative Fuels for Industrial Gas Turbines (Aftur). Applied Thermal Engineering. 24: 1655-1663.
10. Juste GL (2006) Hydrogen Injection as Additional Fuel in Gas Turbine Combustor.Evaluation of Effects.International Journal of Hydrogen Energy. 31: 2112-2121.
11. Jones R, Goldmeer J and Monetti B (2011) Addressing Gas Turbine Fuel Flexibility. GE Energy
12. Haynes J, Janssen J, Russell C and Huffman M (2006) Advanced Combustion Systems for Next Generation Gas Turbines. United States. Dept. of Energy, Washington, D.C.; Oak Ridge, Tenn.
13. Sturgess GJ and Hsu KY (1998) Combustion Characteristics of a Trapped Vortex Combustor.Applied vehicle technology panel symposium.
14. Bruno C and Losurdo M (2007) The Trapped Vortex Combustor: An Advanced Combustion Technology for Aerospace and Gas Turbine Applications. In: Syred N and Khalatov A, Syred N and Khalatov A, editors. Advanced Combustion and Aerothermal Technologies. Springer Netherlands, pp 365-384.
15. Mishra DP (2008) Fundamentals of Combustion, Prentice-Hall Of India Pvt. Limited.
16. Acharya S, Mancilla PC and Chakka P (2001) Performance of a Trapped Vortex Spray Combustor.ASME International Gas Turbine Conference.ASME.
17. Hendricks RC, Shouse DT and Roquemore WM (2005) Water Injected Turbomachinery. NASA, Glenn Research Center
18. Ghenai C, Zbeeb K and Janajreh I (2012) Combustion of Alternative Fuels in Vortex Trapped Combustor. Energy Conversion and Management. In press
19. Ezhil Kumar PK and Mishra DP (2011) Numerical Simulation of Cavity Flow Structure in an Axisymmetric Trapped Vortex Combustor. Aerospace Science and Technology. In Press

20. Xing F, Wang P, Zhang S, Zou J, Zheng Y, Zhang R and Fan W (2012) Experiment and Simulation Study on Lean Blow-out of Trapped Vortex Combustor with Various Aspect Ratios. Aerospace Science and Technology. 18: 48-55.

21. Xing F, Zhang S, Wang P and Fan W (2010) Experimental Investigation of a Single Trapped-Vortex Combustor with a Slight Temperature Raise.Aerospace Science and Technology. 14: 520-525.

22. Bucher J, Edmonds RG, Steele RC, Kendrick DW, Chenevert BC and Malte PC (2003) The Development of a Lean-Premixed Trapped Vortex Combustor. ASME Turbo Expo 2003 Power for Land, Sea, and Air.

23. Dunn-Rankin D (2008) Lean Combustion: Technology and Control, Academic Press, USA.

24. Mosier SA and Pierce RM (1980) Advanced Combustion Systems for Stationary Gas Turbine Engines. Volume I. Review and Preliminary Evaluation.Final Report December 1975-September 1976.pp Medium: X; Size: Pages: 49.

25. Lefebvre AH and Ballal DR (2010) Gas Turbine Combustion: Alternative Fuels and Emissions, Taylor & Francis.

26. Straub DL, Casleton KH, Lewis RE, Sidwell TG, Maloney DJ and Richards GA (2005) Assessment of Rich-Burn, Quick-Mix, Lean-Burn Trapped Vortex Combustor for Stationary Gas Turbines. Journal of engineering for gas turbines and power. 127: 36-41.

27. Cozzi F and Coghe A (2012) Effect of Air Staging on a Coaxial Swirled Natural Gas Flame.Experimental Thermal and Fluid Science. In press

28. Jermakian V, McDonell VG and Samuelsen GS (2012) Experimental Study of the Effects of Elevated Pressure and Temperature on Jet Mixing and Emissions in an Rql Combustor for Stable, Efficient and Low Emissions Gas Turbine Applications. Advanced Power and Energy Program, University of California, Irvine

29. Feitelberg AS, Jackson MR, Lacey MA, Manning KS and Ritter AM (1996) Design and Performance of a Low Btu Fuel Rich-Quench-Lean Gas Turbine Combustor. Advanced coal-fired power systems review meeting. USA DOE, Morgantown Energy Technology Center.

30. Feitelberg AS and Lacey MA (1998) TheGe Rich-Quench-Lean Gas Turbine Combustor. Journal of Engineering for Gas Turbines and Power, Transactions of the ASME. 120: 502-508.

31. Brushwood J (1999) Syngas Combustor for Fluidized Bed Applications 15th Annual Fluidized Bed Conference.

32. Howe GW, Li Z, Shih TI-P and Nguyen HL (1991) Simulation of

Mixing in the Quick Quench Region of a Richburn-Quick Quench Mix-Lean Burn Combustor. 29th Aerospace Sci Meeting.AIAA.

33. Cline MC, Micklow GJ, Yang SL and Nguyen HL (1992) Numerical Analysis of the Flow Fields in aRql Gas Turbine Combustor.
34. Talpallikar MV, Smith CE, Lai MC and Holdeman JD (1992) Cfd Analysis of Jet Mixing in Low NoxFlametube Combustors.Journal of Engineering for Gas Turbines and Power. 114: 416-424.
35. Blomeyer M, Krautkremer B, Hennecke DK and Doerr T (1999) Mixing Zone Optimization of a Rich-Burn/Quick-Mix/Lean-Burn Combustor. Journal of Propulsion and Power 15: 288-303.
36. Wulff A and Hourmouziadis J (1997) Technology Review of Aeroengine Pollutant Emissions. Aerospace Science and Technology. 1: 557-572.
37. Leicher J, Giese A, Görner K, Scherer V and Schulzke T (2011) Developing a Burner System for Low Calorific Gases in Micro Gas Turbines: An Application for Small Scale Decentralized Heat and Power Generation International Gas Union Research Conference.
38. Al-Halbouni A, Flamme M, Giese A, Scherer V, Michalski B and Wünning JG (2004) New Burner Systems with High Fuel Flexibility for Gas Turbines. 2nd International Conference on Industrial Gas Turbine Technologies.
39. Flamme M (2004) New Combustion Systems for Gas Turbines (Ngt). Applied Thermal Engineering. 24: 1551-1559.
40. WEINBERG F (1996) Heat-Recirculating Burners : Principles and Some Recent Developments. Combustion Science and Technology. 121: 3-22.
41. Wünning J (2005) Flameless Oxidation.6th HiTACG Symposium.
42. Cavaliere A and de Joannon M (2004) Mild Combustion.Progress in Energy and Combustion Science. 30: 329-366.
43. Arghode VK, Gupta AK and Bryden KM (2012) High Intensity Colorless Distributed Combustion for Ultra Low Emissions and Enhanced Performance. Applied Energy. 92: 822-830.
44. Li P, Mi J, Dally B, Wang F, Wang L, Liu Z, Chen S and Zheng C (2011) Progress and Recent Trend in Mild Combustion. SCIENCE CHINA Technological Sciences. 54: 255-269.
45. Wang YD, Huang Y, McIlveen-Wright D, McMullan J, Hewitt N, Eames P and Rezvani S (2006) A Techno-Economic Analysis of the Application of Continuous Staged-Combustion and Flameless Oxidation to the Combustor Design in Gas Turbines. Fuel Processing Technology. 87: 727-736.

46. Luckerath R, Meier W and Aigner M (2008) Flox Combustion at High Pressure with Different Fuel Compositions.Journal of Engineering for Gas Turbines and Power.130: 011505.

47. Costa M, Melo M, Sousa J and Levy Y (2009) Experimental Investigation of a Novel Combustor Model for Gas Turbines. Journal of Propulsion and Power 25: 609-617.

48. Levy Y, Sherbaum V and Arfi P (2004) Basic Thermodynamics of Floxcom, the Low-Nox Gas Turbines Adiabatic Combustor. Applied Thermal Engineering. 24: 1593-1605.

49. Lammel O, Schutz H, Schmitz G, Luckerath R, Stohr M, Noll B, Aigner M, Hase M and Krebs W (2010) Flox Combustion at High Power Density and High Flame Temperatures. Journal of Engineering for Gas Turbines and Power. 132: 121503.

50. Arghode VK and Gupta AK (2011) Development of High Intensity Cdc Combustor for Gas Turbine Engines. Applied Energy. 88: 963-973.

51. Khalil AEE and Gupta AK (2011) Distributed Swirl Combustion for Gas Turbine Application. Applied Energy. 88: 4898-4907.

52. Arghode VK and Gupta AK (2010) Effect of Flow Field for Colorless Distributed Combustion (Cdc) for Gas Turbine Combustion. Applied Energy. 87: 1631-1640.

53. Arghode VK, Khalil AEE and Gupta AK (2012) Fuel Dilution and Liquid Fuel Operational Effects on Ultra-High Thermal Intensity Distributed Combustor. Applied Energy. 95: 132-138.

54. Khalil AEE, Arghode VK, Gupta AK and Lee SC (2012) Low Calorific Value Fuelled Distributed Combustion with Swirl for Gas Turbine Applications. Applied Energy. 98: 69-78.

55. Khalil AEE and Gupta AK (2011) Swirling Distributed Combustion for Clean Energy Conversion in Gas Turbine Applications. Applied Energy. 88: 3685-3693.

56. Greenberg SJ, McDougald NK and Arellano LO (2004) Full-Scale Demonstration of Surface-Stabilized Fuel Injectors for Sub-Three Ppm Nox Emissions.ASME Conference Proceedings. 2004: 393-401.

57. Greenberg SJ, McDougald NK, Weakley CK, Kendall RM and Arellano LO (2003) Surface-Stabilized Fuel Injectors with Sub-Three Ppm Nox Emissions for a 5.5 Mw Gas Turbine Engine.International Gas Turbine and Aeroengine Congress and Exhibition.American Society of Mechanical Engineers.

58. Weakley CK, Greenberg SJ, Kendall RM, McDougald NK and Arellano LO (2002) Development of Surface-Stabilized Fuel Injectors with Sub-

Three Ppm Nox Emissions.International Joint Power Generation Conference.American Society of Mechanical Engineers.

59. Cabot G, Vauchelles D, Taupin B and Boukhalfa A (2004) Experimental Study of Lean Premixed Turbulent Combustion in a Scale Gas Turbine Chamber. Experimental Thermal and Fluid Science. 28: 683-690.

60. Mcdougald NK (2005) Development and Demonstration of an Ultra LowNox Combustor for Gas Turbines. USA DOE, Office of Energy Efficiency and Renewable Energy, Washington, D.C; Oak Ridge, Tenn.

61. Arellano LO, Bhattacharya AK, Smith KO, Greenberg SJ and McDougald NK (2006) Development and Demonstration of Engine-Ready Surface-Stabilized Combustion System. ASME Turbo Expo 2006: Power for Land, Sea, and Air.

62. Arellano L, Smith KO, California Energy Commission. Public Interest Energy R and Solar Turbines I (2008) Catalytic Combustor-Fired Industrial Gas Turbine Pier Final Project Report, California Energy Commission, [Sacramento, Calif.].

63. Clark H, Sullivan JD, California Energy Commission. Public Interest Energy R, California Energy Commission. Energy Innovations Small Grant P and Alzeta C (2001) Improved Operational Turndown of an Ultra-Low Emission Gas Turbine Combustor, California Energy Commission, Sacramento, Calif.

64. Arvind G. Rao and Yeshayahou Levy, "A New Combustion Methodology for Low Emission Gas Turbine Engines", 8th HiTACG conference, July 5-8 2010, Poznan.

Chapter 5

PRELIMINARY EXPERIMENTAL STUDY ON PRESSURE LOSS COEFFICIENTS OF EXHAUST MANIFOLD JUNCTION

Xiao-lu Lu,[1] Kun Zhang,[1] Wen-huiWang,[1] Shao-mingWang,[2] and Kang-yao Deng[1]

[1]Education Ministry Key Laboratory for Power Machinery and Engineering, Shanghai Jiaotong University, Shanghai 200240, China

[2]Technology Center of the SAIC Motor, Shanghai 201800, China

ABSTRACT

The flow characteristic of exhaust system has an important impact on inlet boundary of the turbine. In this paper, high speed flow in a diesel exhaust manifold junction was tested and simulated. The pressure loss coefficient of the junction flow was analyzed. The steady experimental results indicated that both of static pressure loss coefficients L_{13} and L_{23} first increased and then decreased with the increase of mass flow ratio of lateral branch and public manifold. The total pressure loss coefficient K_{13} always increased with the increase of mass flow ratio of junctions 1 and 3. The total pressure loss coefficient K_{23} first increased and then decreased with the increase of mass flow ratio of junctions 2 and 3. These pressure loss coefficients of the exhaust pipe junctions can be used in exhaust flow and turbine inlet boundary conditions analysis. In addition, simulating calculation was conducted to analyze the effect of branch angle on total pressure loss coefficient. According to the calculation results, total pressure loss coefficient was almost the same at low mass flow rate of branch manifold 1 but increased with lateral branch angle at high mass flow rate of branch manifold 1.

INTRODUCTION

A very substantial part of the design of a turbo charging system is the exhaust pipe construction. This is because the flow characteristic of the exhaust system has an important impact on the inlet boundary of the turbine, which affects the power out of the turbine. The principle means to understand that the flow characteristics are the numerical simulation. By using a numerical simulation, it is possible to understand the exhaust pipe pressure wave forms and their propagation characteristics and analyze the use of the exhaust energy and cylinder scavenging as well as design and optimization of the exhaust system. In an exhaust system, the pipe is generally a one-dimensional pipe, so a one-dimensional flow analysis can be used. However, there are three dimensional flows in the exhaust manifold junction of the exhaust pipe system; so, using a one-dimensional numerical simulation will result in a large error. Accuracy in the pressure loss coefficient simulation of the exhaust manifold junction of the exhaust pipe system is essential.

During the existing one-dimensional simulation of the engine, when the exhaust gas passes with a higher flow rate, the accuracy of the calculated pressure loss of the tee is relatively low. In papers [1–6], several researchers have studied the pressure loss coefficient of the exhaust manifold junction. The research showed that the pressure loss coefficient is mainly affected by the flow ratio between the public manifold and branch manifold and that 12 circulation types of the pressure loss coefficient calculation formula are deductible. However, during the theoretical derivation process, the junction fluid is assumed to be an incompressible fluid. When the public manifold Mach number is greater than 0.2, the pressure loss coefficients between the calculated and experimental were quite different. In papers [7–9], some experimental research on the pressure loss model of the exhaust manifold junction of the "T" type was done. The research showed that the pressure loss coefficient was related not only to the flow ratio between the public manifold and branch manifold but also to the Mach number in the public manifold. However, during the experiment, the diameter of the branch manifold was only 12 mm, which was much smaller than that in a real engine, and the flow type of the junction was not common in exhaust manifold. In this paper, the high speed flow in a diesel exhaust manifold junction was measured. The pressure loss coefficients of the exhaust manifold junction with high speed flow were analyzed. The results provide reference for the pressure loss junction model considering gas compressibility. At last, the internal flow phenomena of junction were analyzed by CFD model, and the effect of branch angle on total pressure loss coefficient was also studied.

MATERIALS AND EXPERIMENT

A schematic diagram of the experiment is shown in Figure 1. A 400 kW step motor was used to drive the compressor, and the bleed valve was installed behind the compressor to prevent the compressor from surging during the test. The pressure sensor 2, Pitot tube, and temperature sensor 4 were installed in the public manifold that comes after the compressor. Pressure sensor 3 was installed behind the Pitot tube to measure the difference between the total pressure and static pressure. Flow control valves 1 and 2 were the gate valves, and schematic diagram is shown in Figure 2.The flow within the pipe was changed by lifting the valve. A mass flow sensor that could measure range ratio of 1000 : 1 was installed after flow control valve 2 to measure the mass flow of the gas in branch manifold 2. Temperature sensors were installed in the manifold branches 1 to 3 to measure the gas temperature. Pressure sensor 2 was installed in the branch manifold to measure the pressure. Differential pressure transducer 1 was installed to measure the pressure difference between branch manifold 1 and branch manifold 3. Differential pressure transducer 2 was installed to measure the pressure difference between branch manifold 2 and branch manifold 3. To regulate the back pressure in branch manifold 3, a back pressure valve was installed after branch manifold 3. A photograph of the junction is shown in Figure 3. The junction was made by stainless steel to reduce the wall friction loss. The diameter of the three branch manifolds was 66mm, which is equal to the diameter of the pulse turbo charging system of a heavy duty vehicular diesel engine. The angle between branch manifold 1 and 2 was 45o. The point AM was the intersection point of the center lines. The distances between point A and measuring point 1, between point A and measuring point 2, and between point A and measuring point 3 were 260 mm, which was equal to 4 times of the pipe diameter. The lengths of branch manifold 1, branch manifold 2, and branch manifold 3 were all 0.65m.

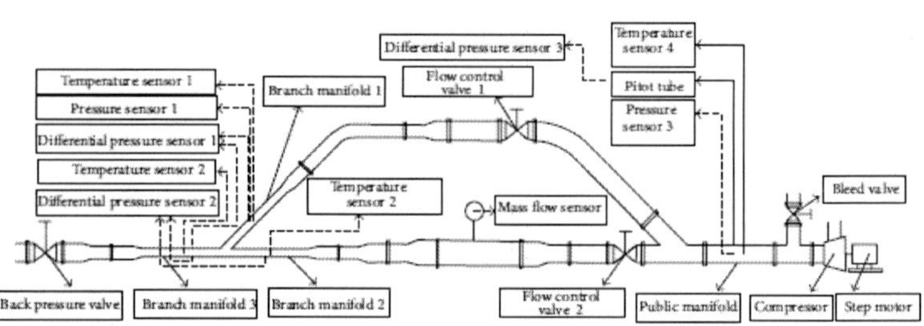

Figure 1: Schematic diagram of the experiment.

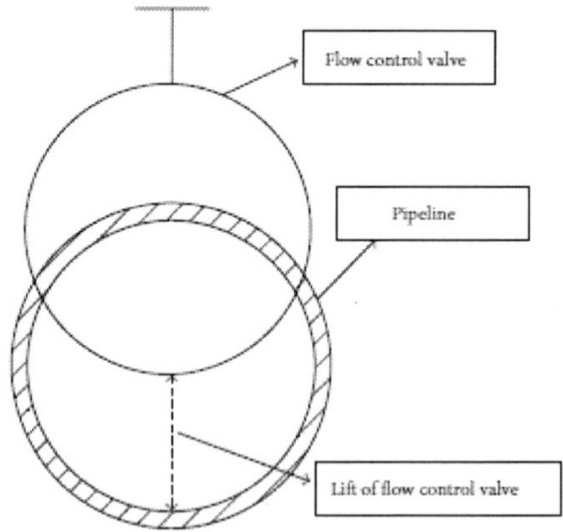

Figure 2: Schematic diagram of the valve.

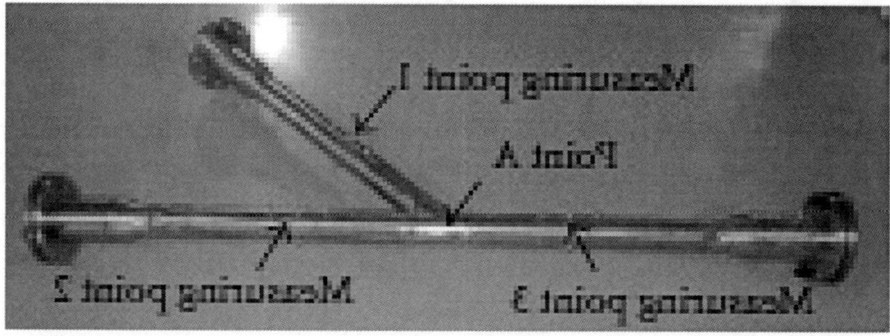

Figure 3: Photograph of the junction.

Measurement ranges of every sensor are shown in Table 1. The output signals of the 10 sensors are the voltage signal, which were collected by aMP426 high-speed data acquisition card. The sampling frequency of the MP426 was 1 KHz. The diameters of the public manifold, flow control valve 1, and flow control valve 2 were 150 mm. The mass flow sensor was installed in a pipe that was 200mmindiameter and2.6m in length. The larger diameter was used to meet the flow measurement range of the compressor, while the measurement accuracy was guaranteed by the long straight pipeline

Table 1. Details of main sensor

Sensor	Model number	Measuring range	Measurement method	Measuring accuracy
Mass flow sensor	GFM200	0–8000 m^3/h (Air standard conditions)	Vortex-shedding flowmeter	±1%
Pressure sensors 1 and 2	NS-I7	0–300 kPa (Gage pressure)	Piezoresistive pressure transducer	±0.5%
Temperature sensors 1, 2, 3, and 4	PT100	0–650°C	Thermocouple temperature sensor	±0.15%
Differential pressure sensors 1 and 2	NS-1151DP	0–50 kPa	Capacitive differential pressure sensors	±0.5%
Differential pressure sensor 3	NS-1151DP	0–30 kPa	Capacitive differential pressure sensors	±0.5%

TEST PROCEDURE AND DATA PROCESSING METHOD

Test Procedure. in the test, in order to prevent a compressor from surging, the purge valve was only opened 30%. By using a step motor, the compressor worked at three speeds: 9000 r/min, 12000 r/min, and 15000 r/min. At each compressor speed, the flow control valve 1 was fully closed first, while flow control valve 2 was fully opened.

The flow control valve 1 was gradually opened, while flow control valve 2 was kept fully opened, Finally, while keeping flow control valve 1 fully opened, close flow control valve 2 gradually from fully opened to fully closed. The valve lift for each test was 5mmwhen the flow control valves were either opened or closed. In each of the situations, output signals from the 10 sensors were collected after the data was stable.

Data Processing Method. Firstly, using the data from pressure sensor 2, temperature sensor 4, and differential pressure sensor 3, calculate the mass flow of the public manifold. Then, using the data from the mass flow sensor under the experimental condition of flow control valve 1 fully closed and flow control valve 2 fully opened, correct the calculated mass flow of the public manifold. The mass flow of branch manifold 3 m_3 was obtained from correcting the mass flow of the public manifold. The mass flow of branch manifold 2 m_2 was obtained from the mass flow sensor. The mass flow of branch manifold 1m_1 was obtained from m_3-m_2.The static pressure of branch manifold 1$P_1$was measured by pressure sensor 1. The static pressure of branch manifold 3 P_3 was calculated from P_1 and the data from differential pressure sensor 1. The static pressure of branch manifold 2 P_2 was calculated from P_3 and the data from differential pressure sensor 2. The density and speed measurement at each branch manifold were ρ_1, ρ_2, ρ_3, $^V{}_1$, $^V{}_2$,and V_3.Thestatic pressure loss coefficient from branch manifold 1 to branch manifold 3L_{13} can be calculated with the following

$$L_{13} = \frac{P_1 - P_3}{P_3}. \qquad (1)$$

The total pressure loss coefficient K_{13} can be calculated with the following equation:

$$K_{13} = \frac{\left(P_1 + (1/2)\rho_1 V_1^2\right) - \left(P_3 + (1/2)\rho_3 V_3^2\right)}{(1/2)\rho_3 V_3^2}. \quad (2)$$

The static pressure loss coefficient from branch manifold 2 to branch manifold 3 L_{23} can be calculated with the following equation:

$$L_{23} = \frac{P_2 - P_3}{P_3}. \quad (3)$$

The total pressure loss coefficient K_{23} can be calculated with the following equation:

$$K_{23} = \frac{\left(P_2 + (1/2)\rho_2 V_2^2\right) - \left(P_3 + (1/2)\rho_3 V_3^2\right)}{(1/2)\rho_3 V_3^2}. \quad (4)$$

EXPERIMENTAL RESULTS AND DISCUSSIONS

Flow Pattern from Branch Manifold 1 to Branch Manifold 3. Figures 4 and 5 show the flow rates in branch manifold 1 and branch manifold 3, respectively, for each working condition of the compressor.

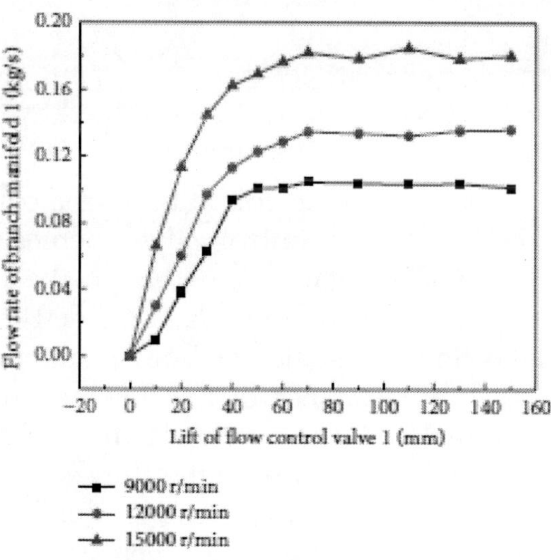

Figure 4: Flow rates in branch manifold 1.

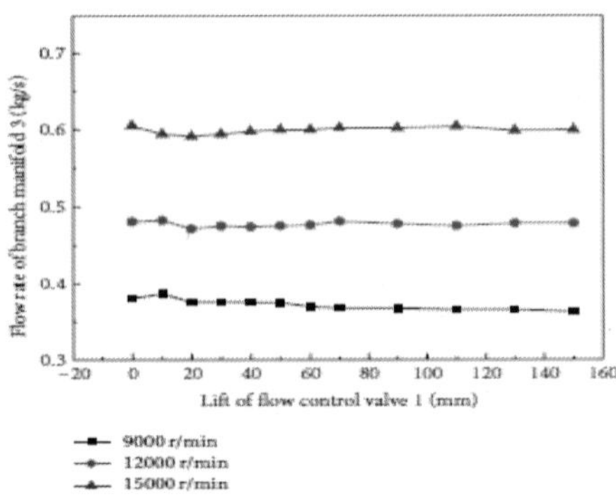

Figure 5: Flow rates in branch manifold 3.

In Figure 4, for each working condition of the compressor, the mass flowin branch manifold 1 initially increased, and then tended to stabilize, with the increase of the flow control valve lift.The mass flow mainly increased in branch manifold 1 when the valve was lifted from 0 to 60 mm. When the valve lift was more than 60 mm, the mass flow m_1 remained unchanged. The reason is that when the valve was initially lifted, the pressure difference between before and after was very big and the mass flow in branch manifold 1 increased as the valve was lifted. Then when the valve lift reached 60mm, the flow area of the valve was similar to the area of branch manifold. After that, when the valve lift continued to open, the mass flow in the branch manifold 1 was no longer determined by the lift of valve 1, but it was mainly decided by its own area.

In Figure 5, for each working condition of the compressor, with the increase of the lift of flow control valve, themes flow in branch manifold 3 was almost unchanged, owing to the flow resistance characteristics of the entire piping system. For the entire piping system, firstly, the flow resistance of branch manifold 1 connects in series with the flow resistance of the flow control valve. Then, they connect in parallel with the flow resistance of branch manifold 2. Finally, they connect in series with the flow resistance of branch manifold 3. When flow control valve 1 is gradually lifted, the flow resistance of the whole piping system decreases, which leads the operating point of compressor moving towards the flow increased. The bleed valve opening has remained at 30%, and the increase in mass flow was mainly dependent on the bleed valve flow and the mass flow in branch manifold 3 changing little. As the compressor speed increases, the mass flow in branch manifold 3 increases, too. Contrasting Figures 4 and 5, when flow control valves 1 and 2 were both opened, the mass flow in branch manifold 1 was less than the mass flow in branch manifold 2.The reason is that the flow resistance of the piping system from the public manifold outlet to branch manifold 1 outlet is greater. When the compressor speed was constant and the mass flow in branch manifold 3 was small, the Mach number in branch manifold 3 changed little. The Mach number in branch manifold 3 was calculated and defined asM_3. According to the pipe system configuration, the Mach number $M3$ was 0.25, 0.34, and 0.45 under compressor speeds of9000 r/min, 12000 r/min and 15000 r/min. M_3 can be used as a substitute for the compressor speed in the following research.

Experimental Pressure Loss from Branch Manifold 1 to Branch Manifold 3

For each working condition of the compressor, the comparison of the pressure difference from branch manifold 1 to branch manifold 3 is shown in Figure 6. The comparison of the static pressure loss coefficient from branch manifold 1 to branch manifold 3, L_{13}, is shown in Figure 7.The comparison of the total pressure loss coefficient from branch manifold 1 to branch manifold 3, K_{13}, is shown in Figure 8.

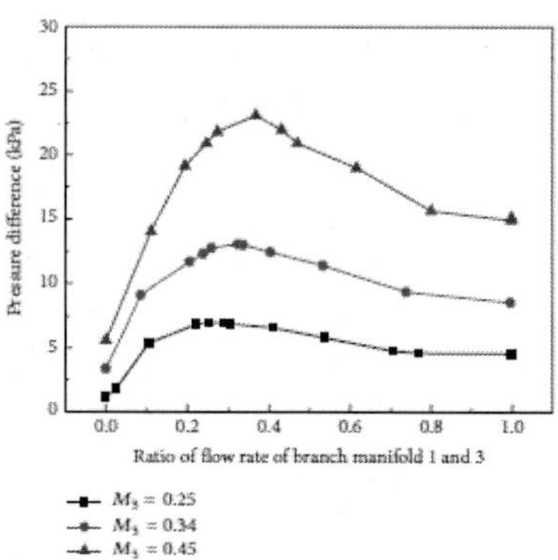

Figure 6: Pressure difference from branch manifold 1 to branch manifold 3.

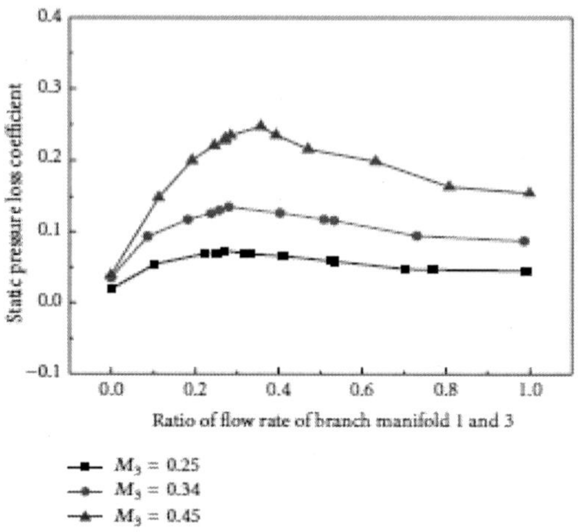

Figure 7. Static pressure loss coefficient from branch manifold 1 to branch manifold 3.

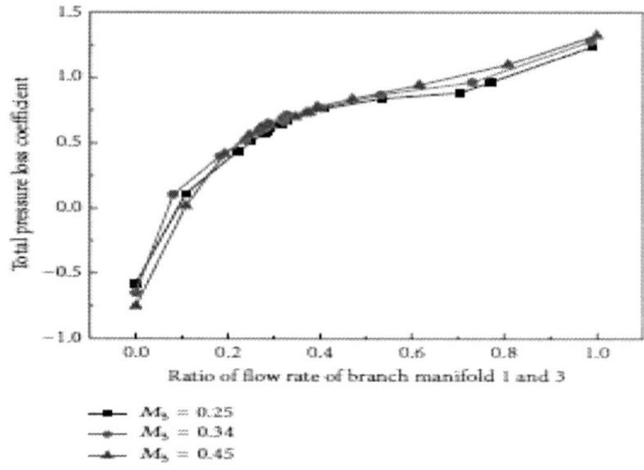

Figure 8. Total pressure loss coefficient from branch manifold 1 to branch manifold 3.

As can be seen from Figure 6, the variation for the three pressure difference curves was consistent. When the Mach Number M_3 was a fixed value, the pressure difference between branch manifold 1 and branch manifold 3 first increased and then decreased with the increase of mass flow ratio of junctions 1 and 3. When the flow rate was constant, them pressure difference between branch manifold 1 and branch manifold 3 increased with the Mach numberM_3. As can be seen from Figure 7, the variation of the three static pressure loss coefficients was the same as the variation for the pressure difference curves in Figure 6. When the Mach number M_3 was a fixed value, the static pressure loss coefficient L_{13} first increased and then decreased with the increase of mass flow ratio of junctions 1 and 3. When the flow ratio was constant, the static pressure loss coefficient L_{13} increased along with an increasing Mach number M_3. As can be seen from Figure 8, the variation of the three total pressure loss coefficients was the same. The total pressure loss coefficient K_{13} always increased with the increase of the flow ratio between branch manifold 1 and branch manifold 3. When the flow ratio between branch manifold 1 and branch manifold 3 was equal to 0, the total pressure loss coefficient K_{13} was negative. The reason for this was that the mass flow of branch manifold 1 was equal to 0 and the air in branch manifold 2 had an extraction effect on the air in branch manifold 1, which leads the pressure in branch manifold 1 to decrease. When the flow ratio was equal to 0,the total pressure loss coefficient K_{13} increased alone with the decrease of the Mach number M_3.This is because the flow rate in branch manifold 3 increased at the same time as the Mach number M_3 and the increased flow rate in branch manifold 3 leads to a greater suction effect on the air in branch manifold 1.When the flow ratio between branch manifold 1 and branch manifold 3 was equal to 1, the higher the Mach number M_3,the higher the total pressure loss coefficient K_{13}.This was due to the greater flow ratio leading to a greater incidence loss, which was a part of the energy loss.

Experimental Pressure Loss from Branch Manifold 2

To Branch Manifold 3. For each working condition of the compressor, a comparison of the pressure difference from branch manifold 2 to branch manifold 3 is shown in Figure9. A comparison of the static pressure loss coefficient L_{23} is shown in Figure 10. A comparison of the total pressure loss coefficient K_{23} is shown in Figure 11.

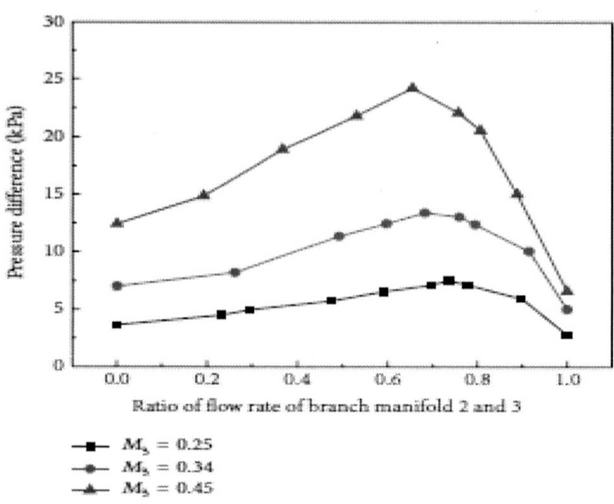

Figure 9. Pressure difference from branch manifold 2 to branch manifold 3.

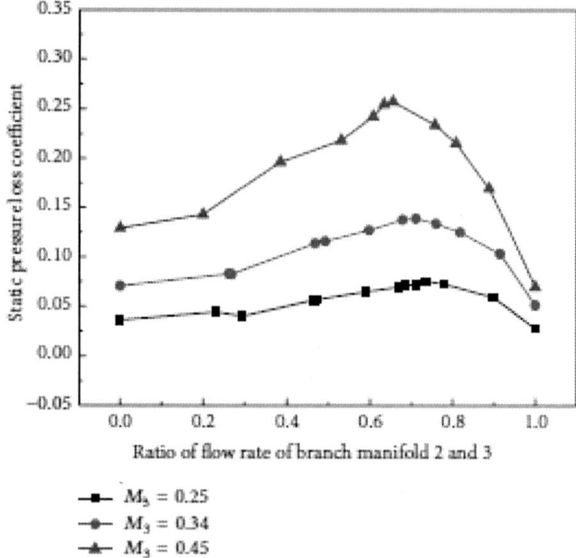

Figure 10. Static pressure loss coefficient frombranch manifold 2 to branch manifold 3.

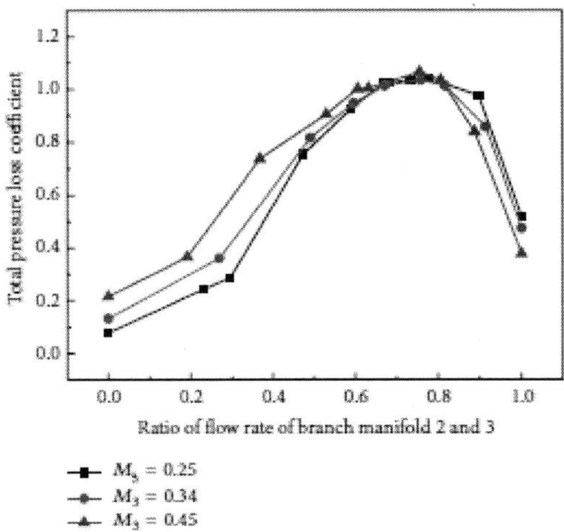

Figure 11: Total pressure loss coefficient from branch manifold 2 to branch manifold 3.

As can be seen from Figure 9, the variation of the three pressure difference curves was consistent. When the Mach number*M*3 was a fixed value, the pressure difference between branch manifold 2 and branch manifold 3 first increased and then decreased as the flow ratio between branch manifold 2 and branch manifold 3 increased. When the flow ratio was constant, the pressure difference between branch manifold 2 and branch manifold 3 increased as did the Mach numberM_3. As can be seen from Figure 10, the variation of the three static pressure loss coefficients was the same as the variation for the pressure difference curves in Figure 9. When the Mach number M_3 was a fixed value, the static pressure loss coefficient L_{23} first increased and then decreased as the flow ratio between branch manifold 2 and branch manifold 3 increased. When the flow ratio was constant, the static pressure loss coefficient L_{23} increased with the Mach numberM_3. As can be seen from Figure 11, the variation of the three total pressure loss coefficients was the same. When the Mach number M_3 was a fixed value, the total pressure loss coefficient K_{23} increased along with the flow ratio between branch manifold 2 and branch manifold 3 increasing first and then decreased when the flow ratio reached 0.7. When the flow ratio between branch manifold 1 and branch manifold 3 was 0, the total pressure loss coefficient

K_{23} increased with the Mach numberM_3.This was due to a greater incidence of loss between branch manifold 1 and branch manifold 3 when the Mach numberM_3 increased, which resulted in a decrease in the total pressure in branch manifold 3. When the flow ratio between branch manifold 2 and branch manifold 3 was equal to 1, the total pressure loss coefficient K_{23} increased with the Mach numberM_3.That was because there was no air flow in branch manifold 1, and the flow rate between branch manifold 2 and branch manifold 3 was the same. The total pressure loss coefficient K_{23} mainly depended on the local pressure loss in branch manifold 1 and its effect on branch manifold 2. Contrast Figure 10 with Figure 11, and it can be determined that the maximum value of the static pressure loss coefficient L_{23} and the total pressure loss coefficient K_{23} are obtained at the same flow ratio 0.7.This was due to the impact loss of flow between branch manifold 1 and branch manifold 2 having the largest effect on the air's energy loss in branch manifold 2 at that moment.

Contrast Figure 7 with Figure 10, and it can be determined that the maximum value of the static pressure loss coefficient L_{13} appeared when the flow ratio between branch manifold 1 and branch manifold 3 was equal to 0.3. In addition, the maximum value of the static pressure loss coefficient L_{23} appeared when the flow ratio between branch manifold 2 and branch manifold 3 was equal to 0.7. The maximum values of L_{13} and L_{23} as well as K_{23}, the total pressure loss coefficient, all appeared at the same condition. Contrast Figure 8 with Figure 11, and it can be determined that when the flow ratio was unchanged, the total pressure loss coefficient K_{23} was strongly influenced by the Mach numberM_3.

JUNCTION SIMULATION AND DISCUSSIONS

45° Junction Simulation. As the flow phenomena in the manifold could not be reflect by experiment, the flow field of junction was simulated by AVL-Fire. The size of simulation model was the same as the measured model which is showed in Figure 12. The inlet boundary conditions of branch manifolds 1 and 2 were mass flow, and the outlet boundary condition of branch manifold 3 was atmospheric pressure.The Mach numbers in branch manifold 3 Q 3 were 0.25, 0.34, and 0.45. For each working condition, the flow ratio between branch manifold 1 and branch manifold 3 Q 13 changed from 0 to 1. The comparison of total pressure loss coefficient between calculated and measured values was showed in Figure 13. The values calculated by CFD were in good agreement with experimental observations. When the flow ratio between branch manifold 1 and branch manifold 3 Q 13 was equal to 0,the junction could be seen as straight pipe. Since the influence of branch manifold 1 can be ignored, the calculated values almost equal to measured values.

FIGURE 12: Simulation model of junction.

Figure 12. Simulation model of junction.

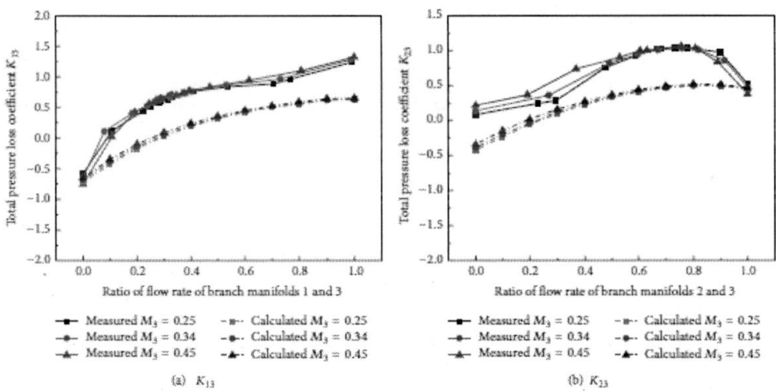

Figure 13. Comparison of total pressure loss coefficient between calculated and measured values.

With the increase of flow rate in branch manifold 1, simulation data were slightly less than the test data, but the change trend of the calculated values were the same as the measured values. Simulation results could reflect the change trend of total pressure loss coefficients K_{13} and K_{23}. As can be seen from Figure 13, the calculated value of total pressure loss coefficient K_{13} always increased when the flow ratio between branch manifold 1 and branch manifold 3 increased, which is the same as the measured value. The calculated value of total pressure loss coefficient K_{23} first increased and then decreased when the flow ratio between branch manifold 2 and branch manifold 3 reached 0.7. By using CFD analysis software, the internal flow phenomenon of junction could be analyzed. Figure 14 shows the velocity fields of junction flow with different mass flow ratio at M_3= 0.34. The mass flow ratio between branch manifold 1 and branch manifold 3 Q_{13} changed from 0 to 1 at the same time when the mass flow ratio between branch manifold 2 and branch manifold 3 Q_{23} changed from 1 to 0. The uniformity of flow at branch joint is increased with the increase of the flow in branch manifold 1.When the mass flow ratio Q_{13} reached 0.6, vortex appeared in the junction and becomes stronger with the increase of Q_{13}.Meanwhile, as can be seen from velocity fields of the junction, the flow impact from branch manifold 1 to branch manifold 2 was enhanced with the increase of Q_{13}. Because of the above reasons, the total pressure loss coefficient K_{13} always increased with the increase of $Q13$. When the mass flow ratio Q_{23} was equal to1, the total pressure loss coefficient K_{23} mainly depended on the local pressure loss in branch manifold 1 and its effect on branch manifold 2. Since the total pressure value of branch manifold 2 decreased with the reduction of flow rate, the total pressure loss coefficient K_{23} decreased alone with the decrease of the flow ratio Q_{23}. The comparison of simulation and test results showed that the following.

1. Simulation could reflect the change trend of total pressure loss coefficient, but it could not predict its value accurately.
2. It is necessary to carry out the experimental study to obtain the accurate value of the total pressure loss coefficient.

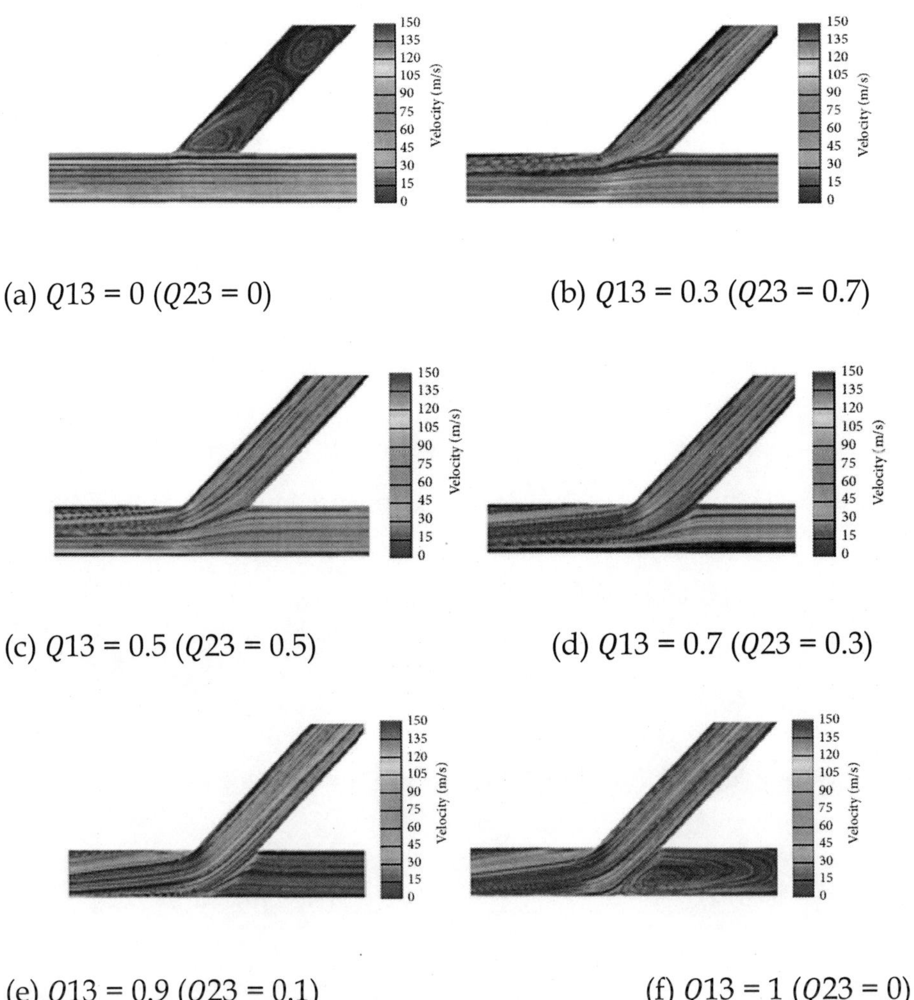

Figure 14. Velocity flow fields of junction with different ratio of flow rate at$M3$ = 0.34.

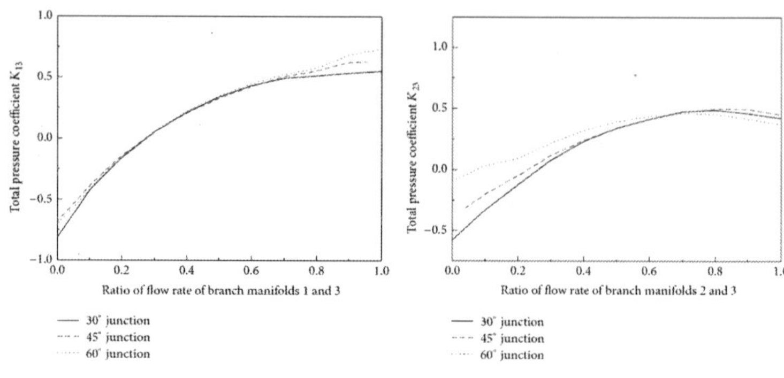

(a) $K13$ (b) $K23$

Figure 15. Comparison of calculated total pressure loss coefficient with various lateral branch angles.

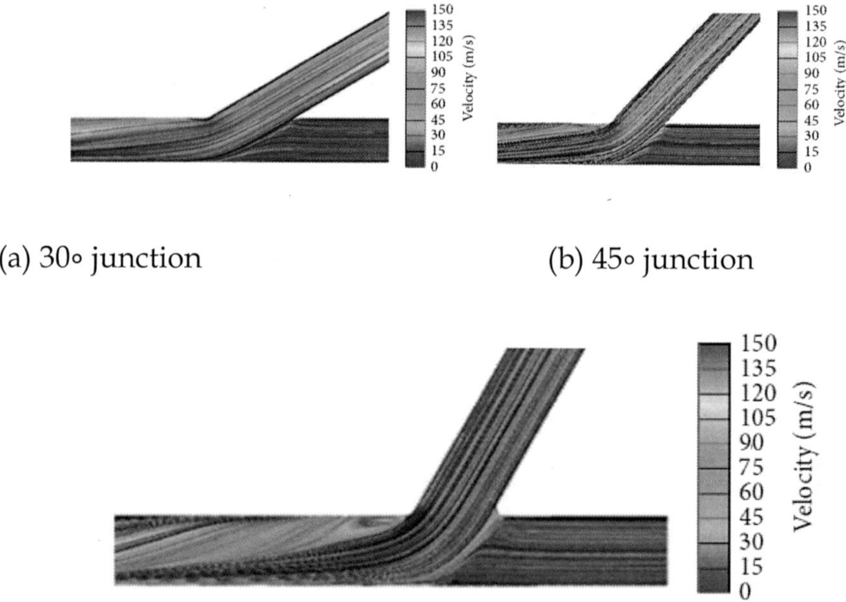

(a) 30∘ junction (b) 45∘ junction

(c) 60∘ junction

Figure 16. Velocity flow fields of junction with different lateral branch angles at Q_{13}=0.9, andM_3= 0.34.

30° and 60° Junction Simulation

As 45° junction is a common type in the exhaust system, the test was based on this type of manifold. But the angle between the branch manifold 1 and the branch manifold 2 may affects the flow condition in the manifold, and other structures also should be studied. As simulation results could reflect the change trend of total pressure loss coefficient, CFD analysis software was used to analyze the effect of branch angle on total pressure loss coefficient. Based on the previous studies, simulation models of 30° and 60° junction were set up. The boundary conditions of the models were the same as 45° junction, and the Mach number in branch manifold 3 $M3$ was 0.34. Figure 15 shows the comparison of calculated total pressure loss coefficient with various lateral branch angles. As can be seen from the figures, when the branch manifold 1 was at low mass flow rate (the branch manifold 1 has less effect on branch manifolds 2 and 3), total pressure loss coefficients of different angles were almost the same. With the increase of mass flow rate of branch manifold 1 and the reduction of mass flow rate of branch manifold 2, the differences between different structures of the manifold are obvious. When the branch manifold 1 was at high mass flow rate, the greater the lateral branch angle, the greater the total pressure loss coefficients $K13$ and $K23$.

Velocity fields of junction flow with different lateral branch angles at $Q13$=0.9 and $M3$= 0.34 are shown in Figure 16. The uniformity of flow at branch joint is increased with the increase of lateral branch angle, and the flow impact between branch manifolds 1 and 2 increased significantly. Meanwhile, vortex appeared in the junction, and it becomes stronger with the increase of the lateral branch angle. For the above reasons, the total pressure loss coefficient$K13$ increased alone with the increase of lateral branch angle.

CONCLUSION

The high speed flow in a diesel exhaust manifold 45° junction was tested, and three junctions with different lateral branch angles were simulated by CFD analysis software. The steady experimental and simulated results indicated that the following.

1. The static pressure loss coefficient *L*13 first increased and then decreased along with the flow ratio between branch manifold 1 and branch manifold 3 increasing; the variation law of the static pressure loss coefficient *L*23 was similar to *L*13.
2. The total pressure loss coefficient *K*13 always increased with the increase of the flow ratio between branch manifold 1 and branch manifold 3; the total pressure loss coefficient *K*23 first increased and then decreased with the increase of the flow ratio between branch manifold 2 and branch manifold 3.
3. The maximum values for the static pressure loss coefficient *L*13, the static pressure loss coefficient *L*23, and the total pressure loss coefficient *K*13 appeared at exactly the same condition.
4. When the flow ratio was unchanged, the total pressure loss coefficient *K*23 was strongly influenced by the Mach number*M*3.
5. Simulation results predicted the change trend of total pressure loss coefficient, but they could not predict their value accurately. It is necessary to carry out the experimental study to obtain the accurate value of the total pressure loss coefficient.
6. When the branch manifold 1 was at low mass flow rate, total pressure loss coefficients of different branch angles were almost the same. At high mass flow rate, the total pressure loss coefficients increased alone with the increase of lateral branch angle.

CONFLICT OF INTERESTS

The authors declare that there is no conflict of interests regarding the publication of this paper.

REFERENCES

1. D. E. Winterbone and R. J. Pearson, Design Techniques for Engine Manifolds—Wave Action Methods for IC Engines, Professional Engineering Publications, London, UK, 1999.
2. D. E. Winterbone and R. J. Pearson, Theory of Engine Manifold Design—Wave Action Methods for IC Engines, Professional Engineering Publications, London, UK, 2000.
3. M. D. Bassett, R. J. Pearson, N. P. Fleming et al., "AMulti-Pipe Junction

Model for One Dimensional Gas-Dynamic Simulations," SAE Paper 2003-01-0370, 2003.

4. M. D. Bassett, D. E.Winterbone, andR. J. Pearson, "Calculation of steady flow pressure loss coefficients for pipe junctions," Proceedings of the Institution ofMechanical Engineers C, vol. 215, no. 8, pp. 861–882, 2001.
5. M. D. Bassett, N. P. Fleming, and R. J. Pearson, "Modelling engines with pulse converted exhaust manifolds using onedimensional techniques," SAE Paper 2000-01-0290, 2000.
6. R. J. Pearson, R. J. Pearson, andD.E.Winterbone, "Estimationof steady flow loss coefficients forpulse converter junctions in exhaust manifolds," in Proceedings of the 6th International Conference on Turbocharging and Air Management Systems, I Mech E Paper No. C554/022/98, I Mech E HQ, London, UK, November1998.
7. J. P´erez-Garc´ıa, E. Sanmiguel-Rojas, J. Hern´andez-Grau, and A. Viedma, "Numerical and experimental investigations on internal compressible flow at T-type junctions," Experimental Thermal and Fluid Science, vol. 31, no. 1, pp. 61–74, 2006.
8. J. P´erez-Garc´ıa, E. Sanmiguel-Rojas, and A. Viedma, "New coefficient to characterize energy losses in compressible flow at T-junctions," AppliedMathematicalModelling, vol. 34, no. 12, pp. 4289–4305, 2010.
9. J. P´erez-Garc´ıa, E. Sanmiguel-Rojas, and A. Viedma, "New experimental correlations to characterize compressible flow losses at 90-degree T-junctions," Experimental Thermal and Fluid Science, vol. 33, no. 2, pp. 261–266, 2009.

Chapter 6

ANALYSIS OF THE UNSTEADY FLOW FIELD IN A CENTRIFUGAL COMPRESSOR FROM PEAK EFFICIENCY TO NEAR STALL WITH FULL ANNULUS SIMULATIONS

Yannick Bousquet,[1,2] Xavier Carbonneau,[2] Guillaume Dufour,[2] Nicolas Binder,[2] and Isabelle Trebinjac[3]

[1] Liebherr-Aerospace Toulouse SAS, 408 avenue des Etats Unis, 31016 Toulouse, France

[2]D´epartement d'A´erodynamique, Energ´etique et Propulsion, ISAE, Universit´e de Toulouse, BP 54032, 31055 Toulouse Cedex 4, France

[3]Laboratoire de M´ecanique des Fluides et d'Acoustique, Ecole Centrale de Lyon, UCB Lyon I, INSA, 36 avenue Guy de Collongue, 69134 Ecully Cedex, France

ABSTRACT

This study concerns a 2.5 pressure ratio centrifugal compressor stage consisting of a splattered enshrouded impeller and a vaned diffuser.

The aim of this paper is to investigate the modifications of the flow structure when the operating point moves from peak efficiency to near stall. The investigations are based on the results of unsteady three-dimensional simulations, in a calculation domain comprising all the blade. A detailed analysis is given in the impeller inducer and in the vaned diffuser entry region through time-averaged and unsteady flow field. In the impeller inducer, this study demonstrates that the mass flow reduction from peak efficiency to near stall leads to intensification of the secondary flow effects. The low momentum fluid accumulated near the shroud interacts with the main flow through a shear layer zone. At near stall condition, the interface between the two flow structures becomes unstable leading to vortices development. In the diffuser entry region, by reducing the mass flow, the high incidence angle from the impeller exit induces a separation on the diffuser vane suction side. At near stall operating point, vortices from the separation is shed into vortex cores which are periodically formed and convicted downstream along the suction side.

INTRODUCTION

Centrifugal compressors for the aeronautical field are expected to achieve high pressure ratios and high efficiencies at design operating point while minimizing the element size. In this context, typical centrifugal compressor stages are composed of high speed impellers with vaned diffusers to achieve the high pressure recovery in a reduced space. On the other hand, extending the operating range as much as possible is also an important design constraint. As for axial configurations, the limitation at low mass flow rates comes from the rotating stall and/or surge phenomena. Rotating stall is characterized by the presence of one or several cells rotating around the annulus, either in the impeller or in the diffuser. Surge is a system dependant phenomenon associated to large amplitude oscillations of the pressure through the compressor system [1]. The works of Galindo et al. [2] show that the surge intensity can be modified by changing the length of the duct downstream the compressor. However, operating the system in these unstable conditions induces a dramatic drop of performance associated with mechanical stresses that may cause the failure of the compressor. Therefore, a margin (surge margin) is taken to keep away the operating point from these phenomena,

leading to an operating range reduction. Backswept impellers are widely used to increase the stability of the impeller and finally the stability of the stage [3]. Efforts have also been dedicated to identify flow control strategies in order to delay the emergence of instability. Different techniques are presented by Skoch [4] in a high speed centrifugal compressor. However, the use of stabilization techniques to extend the operating range requires a good comprehension of flow mechanisms occurring before the stall diffuser.

Stall phenomenon in compression systems has been studied for almost sixty years. The experimental works of Emmons et al. [5] and Mizuky and Oosawa [6] in a low pressure ratio centrifugal impeller with a vaneless diffuser show that rotating stall occurs in the inducer and leads to surge. In a high pressure ratio centrifugal compressor with vaned diffuser, Wernet et al. [7] and Tr´ebinjac et al. [8] observed separation of the boundary layer on the diffuser vane before the surge inception.

Before the onset of these developed instabilities, two families of precursors have been observed, firstly in axial configuration [9] and then in centrifugal configuration [10]. The first precursor signal is the growth of a small amplitude disturbance with a long length scale referenced as modal stall, while the second concerns the growth of a higher amplitude disturbance but with smaller length scale (several blade passages) termed spike. Moreover, most of the past works deal either with high pressure ratio (>5) transonic compressors using vaned diffuser or low pressure ratio (<2) compressors using vaneless diffuser. For the first category, there is evidence that the diffuser entry region is often responsible for the surge onset while for the second impeller inducer seems to be the weakest zone in terms of flow instability. The present study concerns a subsonic and moderate pressure ratio (2.5) centrifugal compressor stage composed of a splattered impeller and a vaned diffuser.

In addition, work on the topic comes generally from experimental investigations where the flow description is therefore partial at best. In this study, investigations are performed through a computational fluid analysis. Near the surge line, unsteady phenomena uncorrelated to the blade passing frequency may appear. Therefore, to capture all of the spatial and temporal content, the calculation domain has been extended to all the blade passages leading to 60-millionpointmesh. Since, the surge onset is generally triggered in the impeller inducer or in the vaned diffuser entry zone, efforts are concentrated at these locations. The two main objectives of this paper are to (i) numerically analyze the modifications of the flow structure induced by the mass flow reduction from peak efficiency to a near stall condition and (ii) investigate in depth through unsteady flow data sets the near stall operating point in order to propose different scenarios which may lead to flow breakdown. In the first part of the paper, the study case and the numerical procedure are presented. Then, the numerical model is validated thanks to experimental measurements. Afterwards, detailed analysis of the unsteady flow features in the impeller inducer and finally in the diffuser entry zone is given depending on the operating point. The last two parts constitute the scope of the paper.

TEST CASE PRESENTATION

The centrifugal compressor stage considered for this study has been designed by Liebherr-Aerospace Toulouse SAS and is integrated in an air-conditioning system. The stage is composed of a backswept splattered enshrouded impeller, a vaned diffuser, and a volute. The design specification is based on a stage static-to-total pressure ratio of 2.5 with a design rotation speed of 38000 rpm. The impeller contains 8 main blades and 8 splitter blades with a back sweep angle of 32 ∘. The impeller exit radius is about 100 mm. The vaned diffuser consists of 21 wedge blades (see Figure 1).

NUMERICAL PROCEDURE

Flow Solver

Computations are Performed using the *elsA* solver developed by ONERA and CERFACS [11]. It solves the three-dimensional unsteady compressible Reynolds averaged Navier-Stokes equations, based on a cell-centered finite volume approach on multiblock structured grids. The turbulent viscosity is computed with the one-equation Spalart-Allmaras model [12]. The convective fluxes are computed with the centered second-order scheme with artificial dissipation of Jameson and diffusive fluxes with a second order centered scheme.The time-marching integration is performed with an implicit scheme composed of the backward Euler scheme and a scalar lower-upper symmetric successive overrelaxation method (LUSSOR) proposed by Yoon and Jameson [13]. This time-marching scheme is coupled with a second-order dual time stepping method proposed by Jameson [14]. The number of physical time steps to discretize a complete rotation is set to 1680, corresponding to 210 time steps per impeller main blade passing (impeller has 8 main blades). This discretization value permits a correct description of the impeller-diffuser interactions according to what is usually reported in the literature [15]. For each physical time step, iterations are performed in the inner loop until two orders of magnitude residual reduction are reached. This condition is satisfied in less than 10 sub iterations. To reach the periodic state, at least 12 impeller rotations are needed, equivalent to approximately 20000 physical time steps. They are performed using 512 computing cores and require 200000 CPU hours for a single operating point.

Boundary Conditions

The computational domain contains the centrifugal impeller with the inlet bulb, the vaneless space, and the vaned diffuser and ends with a 90° turning pipe used as a computational buffer zone to damp possible reflections from the exit boundary condition (Figure 2). The computational domain extends 1.5 impeller inlet tip radius upstream and downstream of the blade rows. As the meshing of the complete annulus requires significant CPU cost, the volute is not taken into consideration. Considering the inlet boundary conditions, the total pressure, the total temperature, and the flow angles (axial flow) are prescribed. When the operating point moves toward the surge line, the slope of the stage pressure ratio characteristic may reach zero or even positive values. Therefore, a static pressure exit condition is not adapted because two solutions may be obtained (different mass flow rates) for a same outlet pressure while a fixed mass flow condition may lead to numerical problem.

Figure 1. 3D sketch of the compressor stage hub with the return downstream channel.

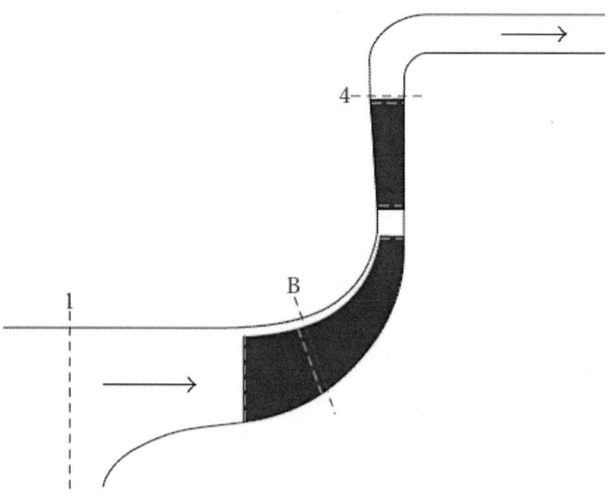

Figure 2: Meridional view of the calculation domain.

Therefore, in the present study, the outlet is modeled using a throttle condition, coupled with a simplified radial equilibrium law. The outlet static pressure p_{out} is set by the following relation:

$$p_{out}(n+1) = P_{i0} + \lambda\left(\frac{\dot{m}(n)}{m_{ref}}\right)^2, \qquad (1)$$

where P_{i0} is the inlet total pressure, m□ (n) is the mass flow rate at iteration n through the exit, section and λ is the throttle parameter. The simulated operating point can move from choked point to surge line by simply increasing the value of the throttle parameter. The rotor-stator interface is treated with a sliding mesh method. In elsA, the communication through the sliding surface is performed using a distribution of fluxes. This approach rigorously ensures conservatively for planar interfaces, which is almost the case here. Details on the implementation and use of the sliding mesh technique with the elsA code can be found in the work of Filola et al. [16] and Gourdain et al. [17]. At the blade, hub and shroud walls no-slip adiabatic conditions are prescribed.

Mesh Parameters. The structured mesh grid was generated with Autogrid V5 using classical H, O, and C topologies. In order to obtain mesh independent results, the parameters to generate the mesh result from a previous study [18] performed in the same configuration. The size of the first cell is set to 3 μm corresponding to a normalized wall distance $y+$ well below 3 at the walls. The impeller main blade passage grid and splitter blade passage grid consist of 89 points in the span wise direction including 29 points in the gap region, 92 points in the pitch wise direction, and 161 points in the stream wise direction. The diffuser blade passage contains 57 points in the span wise direction, 119 points in the pitch wise direction, and 141 points in the stream wise direction. The impeller blade passage and the diffuser blade passage include, respectively, $2.6{*}10^6$ and $1.7{*}10^6$ cell grid. The single passage is repeated to obtain the full annulus and the calculation domain reaches a total of approximately 60 million points.

Data Extraction. Unsteady computations are performed for three operating pointsOP1 (peak efficiency),OP2, and NS (near stall). As will be illustrated later, unsteady fluctuations for the OP1 and OP2 are only generated by impeller-diffuser interactions. In other terms, for these two operating points, the flow is time periodic in the frame of reference of each row. After reaching the unsteady periodic state, a full rotation of the rotor is performed to extract data (unsteady and time averaged). The time-averaging period is equal to one rotor rotation. For the NS operating point, unsteady effects are not only limited to rotor-stator interactions, and the natural periodicity of the flow is no longer valuable. Therefore, after reaching the stable state, the simulation has been extended during six rotor rotations to validate the stability of the operating point and to extract data. The time-averaging period is therefore equal to 6 rotor rotations. Because of the domain size, the data extraction of the complete flow field is hardly affordable with a correct temporal resolution. Therefore, data extraction is segregated into local information recorded at each time step, two-dimensional planes extracted every 10 time steps, and the complete three-dimensional field saved three times per impeller rotation.

Numerical Model Validation

Figure 3 depicts the total-to-static pressure ratio defined as $\pi = p_4/p_{01}$ as a function of the corrected mass flow from numerical simulations results and measurements, for the design speed line.The corrected mass flow is defined as

$$\dot{m}_{\text{cor}} = \frac{\dot{m}\sqrt{T_{01}/T_{\text{ref}}}}{p_{01}/p_{\text{ref}}}. \tag{2}$$

The experimental value of p_4 is the mean value of three static pressure probes located on the hub surface at diffuser exit radius (plane 4, see Figure 2).The static pressure from numerical results is extracted at the same location. All the computed operating points show good agreement with experimental data. The main objective of the measurements was to validate the numerical model. In the following sections, investigations will only consider the numerical simulation results.

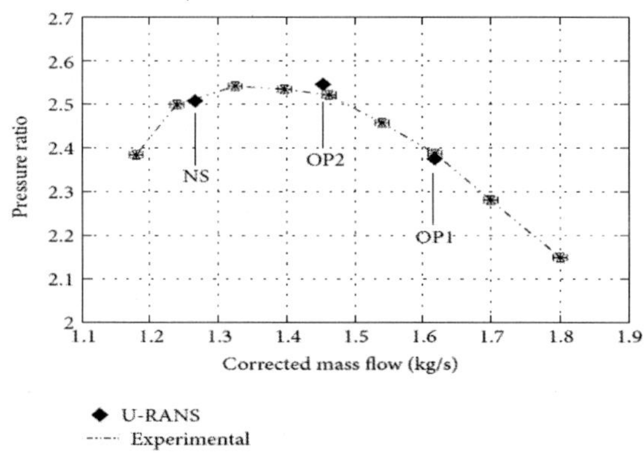

Figure 3. Pressure ratio of the compressor stage.

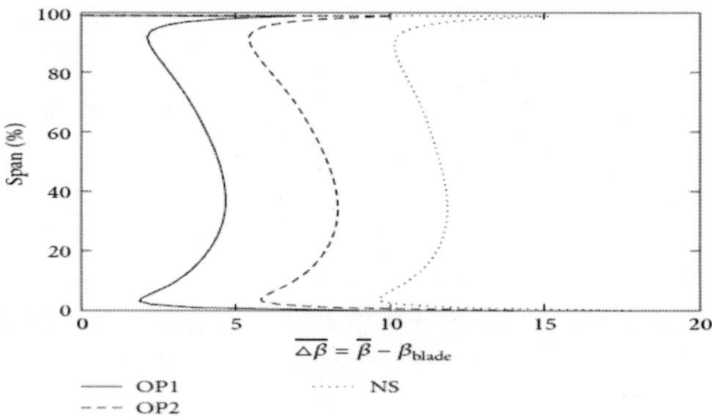

Figure 4. Time-averaged ax symmetric profile from hub to shroud of the incidence angle at the impeller inlet.

IMPELLER FLOW STRUCTURE ANALYSIS

When the centrifugal impeller is responsible for the surge onset, the origin of the destabilization occurs generally in the inducer and near the shroud. The unsteady numerical results at this specific location are now presented.

Secondary Flow Effects

The centrifugal impeller flow structure has been studied continually since the 1970s. The work of Eckardt [19, 20] shows that the main flow is affected by secondary flows, responsible of the jet-wake structure classically observed in centrifugal impellers. They are produced by reorientation of transverse vortices (in the boundary layers) into longitudinal vortices under the effects of curvature and rotation [21]. The meridional curvature part effect drives the low momentum fluid in the boundary layer of the blades (radial migration from hub to shroud), while the rotation effects act mainly on the fluid particles in the hub and shroud boundary layer (migration from PS to SS). The secondary flow effects are generally noticeable from the axialradial turn and intensify up to the impeller exit due to the thickening of the boundary layers. To investigate the impeller flow structure depending on the operating points, analysis of the inlet flow field conditions is first conducted.

The operating point displacement frompeak efficiency to near stall leads to a gradual decrease of the mean meridional velocity. Since the rotation speed is constant, the incidence angle on the impeller main blades rises. In the study case, the mass flow reduction induces an increase of the incidence angle of approximately 4° fromOP1 to OP2 and 4° again from OP2 to NS (Figure 4).Because of the high incidence at low mass flow rate and of the pressure rise occurring in the impeller (adverse pressure gradient), the boundary layer thickens and separates on the impeller main blade suction side. Figure 5(a) shows the reduced axial velocity profile in the suction side boundary layer, at 50% of the span near the leading edge for the three operating points. At high mass flow rate (OP1), the incidence angle does not exceed the critical value and no separation is observed. By moving to OP2, the critical incidence angle is reached and a small separation occurs while at NS condition it significantly increases. For OP1, the boundary layer thickness represents 1% of the pitch wise. By moving to OP2, the boundary layer thickness is twice larger while at NS it is four times larger. As a consequence of the boundary layer thickening at low mass flow rate, the secondary flow effects become stronger and are noticeable from the leading edge of the impeller blade. Figure 5(b) plots the reduced radial velocity profile in the suction side boundary layer, at 50% span near the leading edge, for the three operating points. At design condition, the boundary layer is not enough developed (boundary layer is too thin) to induce a radial migration. Therefore, an increase of radial velocity near the suction side of the impeller blade is not observed. At low mass flow rates, due to the thickening of the boundary layer, the secondary flow effects can clearly be observed by the significant rise of the radial velocity near the suction side of the impeller blade. Moving from OP2 to NS leads to intensification of this mechanism and the increase of the radial velocity is larger.

The low momentum fluid near the main blade suction side is then transported along the blade from hub to shroud (positive radial velocity). At the tip of the blade, it is transported and stretched by the leakage flow of the main blade in the middle of the channel (Figure 6). As a result, by observing the time-averaged meridional velocity at section B (Figure 7), at low mass flow rates, a velocity deficit region (wake) can be noticed in the right channel close to the shroud. This region results from the combination of secondary and leakage flows. By reducing the mass flow along the speed line fromOP2 toNS, the wake region significantly enlarges.

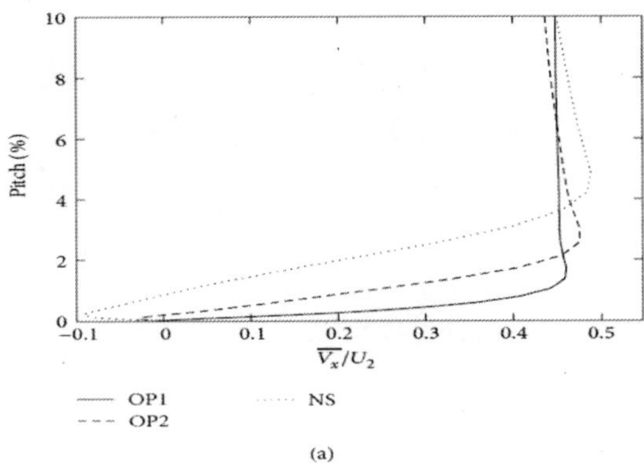

(a)

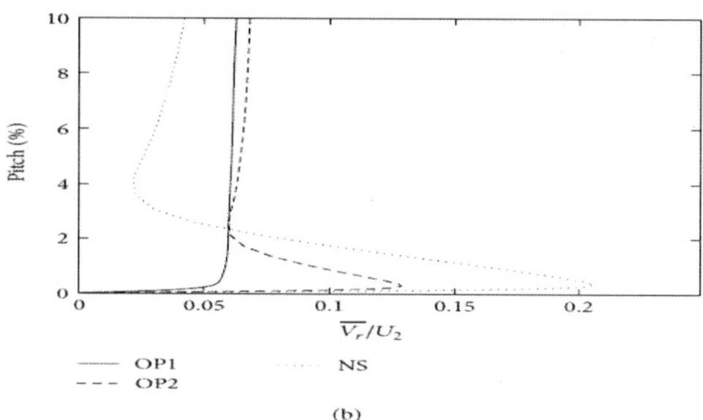

Figure 5. Time-averaged reduced axial velocity (a) and time-averaged reduced radial velocity (b), in the main blade suction side boundary layer, at 50% span, 2mm downstream the leading edge.

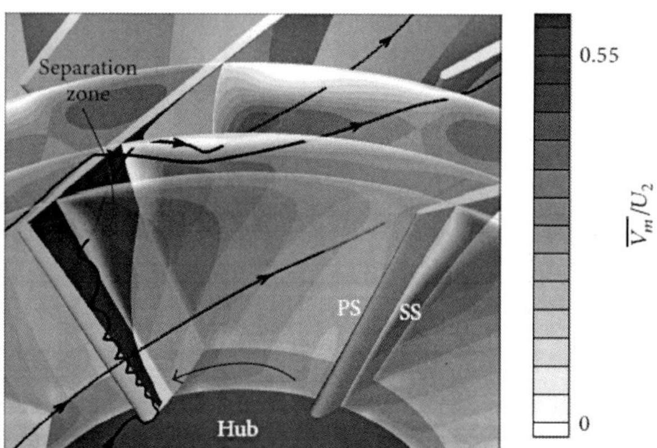

Figure 6. Time-averaged reduced meridional velocity contours for the near stall operating point. left channel, the mass flow reduction does not affect the flow structure which is principally composed of the core flow.

Unsteady Flow Analysis. To investigate the unsteady flow pattern for the simulated operating points, a spectral analysis is performed from a local extraction, recorded at each time step leading to an approximate sample frequency of 1 Mhz. In normal operating conditions (stable conditions, constant rotation speed. . .), unsteady phenomena are mainly induced by the blade passing effects. Therefore in the impeller, the flow is time periodic with a period of $T_R= 2\pi/\Omega rNs$ while in the diffuser the period is $Ts= 2\pi/\Omega rNr$. This specificity is often used to reduce the calculation domain to one single blade passage. Afterwards, the solution can be obtained with an unsteady calculation model using a spatial temporal periodicity (chorochronic approach).

Figure 8 plots the frequency spectra of a static pressure probe linked to the relative frame located at 90% span, at the impeller inlet. It can be seen that the frequency content for OP1 and OP2 is limited to the blade passing frequency located at $f*= 21$ (diffuser has 21 vanes). Since the compressor operates in subsonic conditions, potential effects from the vane diffuser can propagate upstream and reach the impeller inlet. These operating points could have been simulated using a spatial-temporal periodicity. However, at NS condition, the frequency content is not only limited to the blade passing frequency and a periodic unsteady phenomenon emerges at$f*= 6$. This occurrence shows clearly the need of meshing all the blade passages to the analyzed near stall operating point. The following part focuses on a detail analysis of this frequency emergence occurring at NS.

As discussed in the previous section, the impeller flow structure is composed of the main flow and of the secondary and leakage flows, leading to high and low meridional velocity zones (see Figure 7).The interface between these two flow structures is a region of significant shear. By reducing the mass flow, the meridional velocity deficit due to leakage and secondary flows increases and at NS condition it is such that the velocity gradient is enough to create an interface instability. Vortices are formed at the interface and are transported with the main flow downstream. Figure 9 depicts the instantaneous meridional velocity and instantaneous streamlines at 90% span, in the relative frame.The incoming flow region can be identified as the high velocity zone (blue region) while secondary and leakage flows are marked with a significant low velocity zone (white region). At $t = 9.1Tr$, a vortex is seen near the splitter blade leading edge while the black point represents the location of the vortex formation. The vortex is then transported by the main flow and grows. At $t = 10T_R$ the vortex is generated and two vortices can be noticed. During one rotor rotation, six vortices are formed and shed, responsible of the fundamental frequency$f*$= 6 seen in the frequency spectra (Figure 8). All the blade channels create the same phenomenon, but the vortices location varies slightly between the different passages.

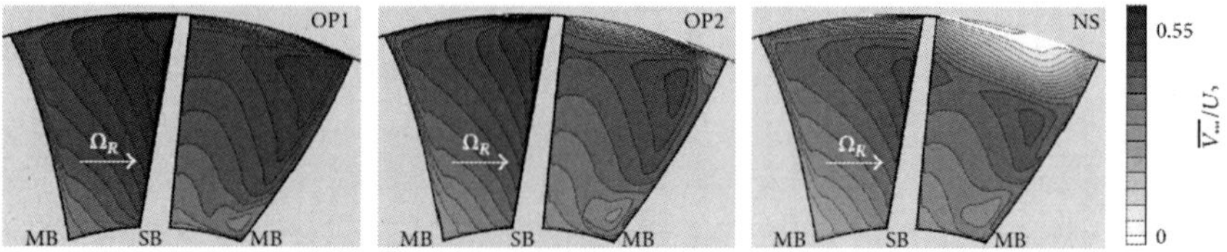

Figure 7. Time-averaged reduced meridional velocity contours at section B, for the three operating points.

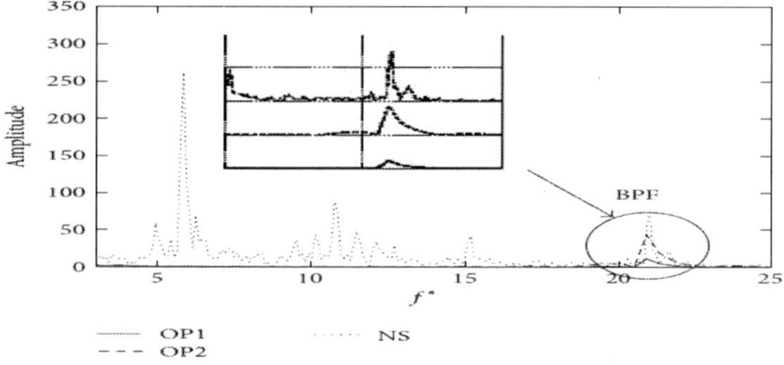

Figure 8. Frequency spectra of a pressure signal in the relative frame, at 90% span, at the impeller inlet.

Similarities with Axial Compressors. In axial compressors, M¨arz et al. [22] have also observed vortices in the rotor tip region, induced by the interaction between the reverse flow near the trailing edge, the tip clearance flow, and the incoming flow. The vortices move from the suction side to the pressure side and are referenced as rotating instability. Due to the reverse flow near the trailing edge, the vortices do not move downstream and travel circumferentially. The compressor operates in a stable mode even with this rotating instability. In the present case, as there is no reverse flow close to the shroud, the vortices are formed and convicted downstream. According to Duc Vo et al. [23], there are two necessary conditions for the spike disturbance formation in axial compressor. The first one is that the interface between the incoming flow and the tip clearance flow becomes parallel to the leading edge plane, permitting the tip clearance flow to spill into the following blade passage.The second is the initiation of reverse flowat the trailing edge.

The first criterion has been investigated in our case by observing the interface position. Figure 10 shows the time-averaged entropy contour map at the tip of the blade. The interface position depends on the flow momentum balance between the incoming flow and the leakage flow. The mass flow reduction induces a decrease of the incoming flow momentum and an increase of the leakage flow momentum due to the blade loading increase. As a consequence, when the mass flow is reduced, the interface between the two flow structures becomes more tangential. For OP1 and OP2, the interface line (red line in Figure 10) goes from the main blade leading edge to the splitter blade leading edge. AtNS condition, the interface line is significantly displaced but does not reach the leading edge of the adjacent blade. Therefore, we hypothesize that if the mass flow is further reduced, the interface between the incoming and leakage flows will become parallel to the leading edge plane, permitting the leakage flow to spill into the following passage and leading to impeller rotating stall.

DIFFUSER INLET FLOW STRUCTURE ANALYSIS

For the present compressor stage, the diffusion process is accentuated through a vaned diffuser. As already mentioned, in the literature concerning configurations with vaned diffusers, there is evidence that surge is often triggered in the vaneless space or in the semivaneless space.This part focuses on the description of the flow at this location.

Mass Flow Reduction Effects. At the impeller exit, when the operating point is displaced to the surge region along the speed line, the mean tangential velocity increases due to the rise of work input while the mean radial velocity decreases. Therefore, the vanes of the diffuser have to operate with a higher incidence angle. In addition, all along the meridional curvature, secondary and leakage flows continue to interact with the main flow inducing ahighly distorted pattern in both the span wise and the pitch wise directions. The present part investigates the impeller exit flow conditions depending on the three operating points. Since the work of Dean and Senoo [24], the impeller exit flow structure is described in the pitch wise direction by considering the jet-wake model. The wake zone contains low radial velocity and high absolute tangential velocity flows inducing high absolute flow angle values.

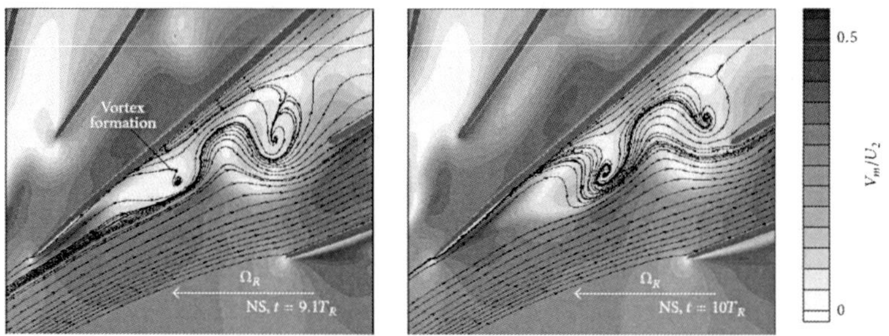

Figure 9. Instantaneous reduced meridional velocity contours and streamlines at 90% span in the inducer.

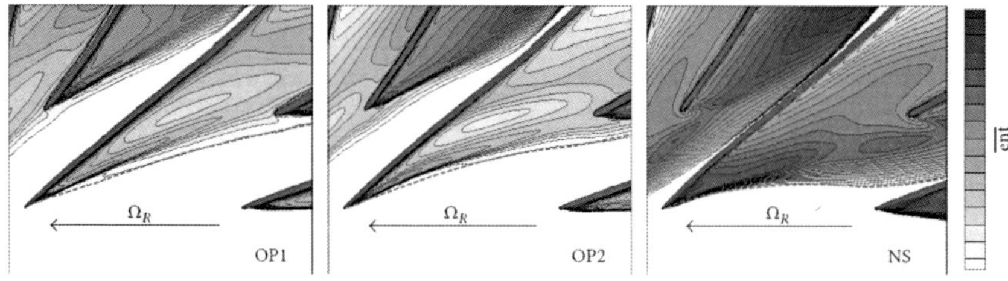

Figure 10. Time-averaged entropy contours at the blade tip (98% span) in the inducer for the three operating points.

Figure 11(a) shows the time-averaged pitch wise distribution of the flow angle profile at 90% span. Considering the three operating points, major differences occur in the right channel (from 50% to 100% of the pitch). The increase of flow angle values at OP2 and NS is linked to the low meridional velocity zone extent at low mass flow rate, noticed in the impeller inducer near the shroud (see Figure 7). Due to the impeller rotation, this pitch wise distortion induces temporal fluctuations of the flow angle at the diffuser inlet, which play a role in the unsteady flow structure described in the following. Besides this pitch wise distortion, the study of Deniz et al. [25] leads to the conclusion that the diffuser performance is mainly determined by the ax symmetric time-averaged inlet flow angle. Figure 11(b) shows this quantity for the three operating points. Diffuser vane angle (metal angle) is constant from hub to shroud. Due to shroud and hub curvatures, the flow is decelerated and the boundary layer thickness increases on the convex shroud surface although a transfer of flow toward the concave hub side induces a radial velocity increase. Therefore, from hub to 85% span and for the three operating points, the flow angle naturally rises while above 85% span it rudely increases until the shroud due again to secondary flows and leakage flows effects. The compressor stability may be affected when the incidence angle becomes positive ($\alpha - \alpha_{blade} > 0$), leading to the possibility of boundary layer separation on the diffuser vane suction side. As the mass flow is reduced, the part of the span in that situation significantly extends from 70–100% of the span to 40–100% of the span (see Figure 11(b)). In addition to the high incidence angle value near the shroud, the pressure gradient in the semivaneless space increases by reducing the mass flow. Therefore, given these conditions, a boundary layer separation occurs on the diffuser vane suction side for OP2 and NS. Figure 12 illustrates the separation zone by representing the time-averaged radial velocity contour and the streamlines. The red zone shows negative radial velocity (reverse flow) while the blue zone shows positive radial velocity. As the flow is subsonic in the vaneless space, the separation bubble yields to a flow deviation, to a rise of flow angle, and finally to a negative radial velocity zone. The mass flow reduction from OP2 to NS induces an enlargement and a displacement of the separation bubble along the suction side toward the leading edge, yielding to a more intense reverse radial flow.

Near Stall Condition Analysis. The unsteady flow structure in the boundary layer separation shown previously has been analyzed with the λ_2 vortex criterion [26]. The velocity gradient tensor J is decomposed into its symmetric part S and antisymmetric part$\Omega$and the eigenvalues of $J^2+\Omega^2$ are determined. If the second eigenvalue

λ_2 is negative in a region, it belongs to a vortex core. This criterion is applied in the vaned diffuser considering the three-dimensional unsteady data.

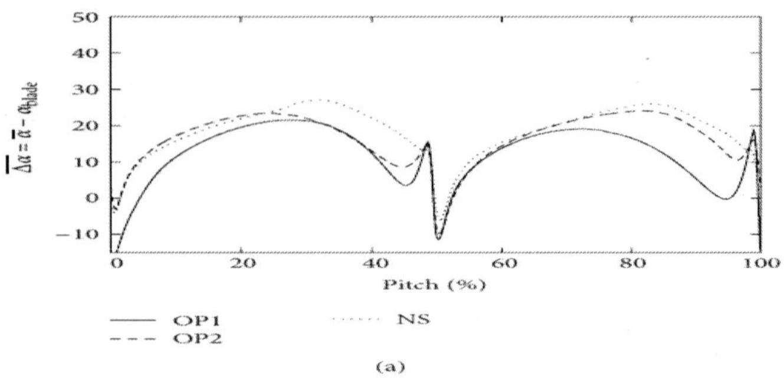

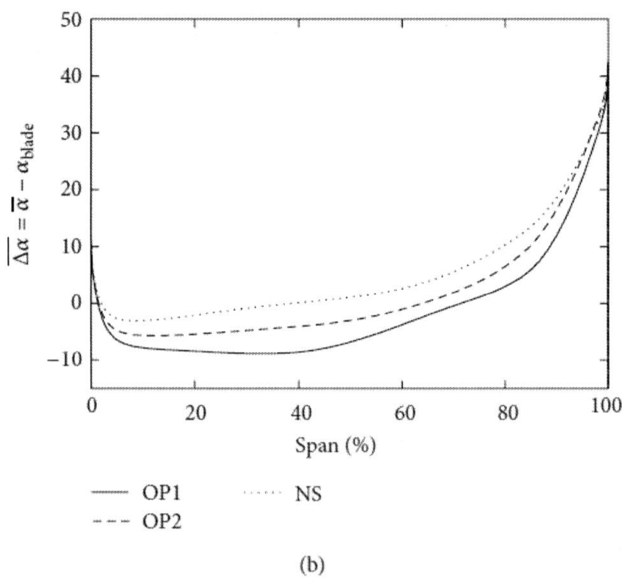

Figure 11. Time-averaged pitchwise distribution at 90% span (a) and axisymmetric profile (b) of the incidence angle, 1mm after the impeller exit.

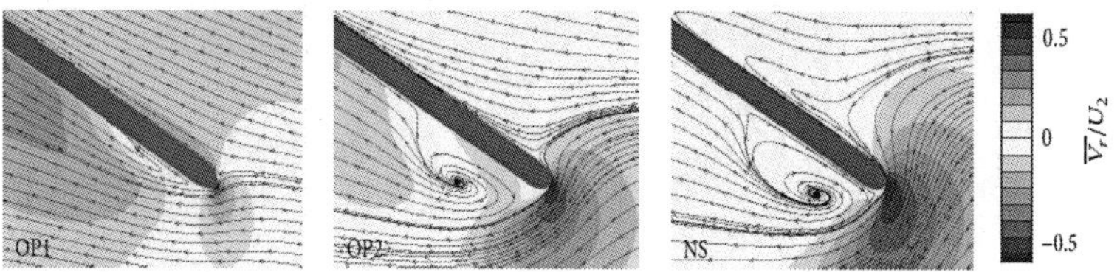

Figure 12. Time-averaged reduced radial velocity contours and streamlines, at 90% span and near the vane leading edge, for the three operating points.

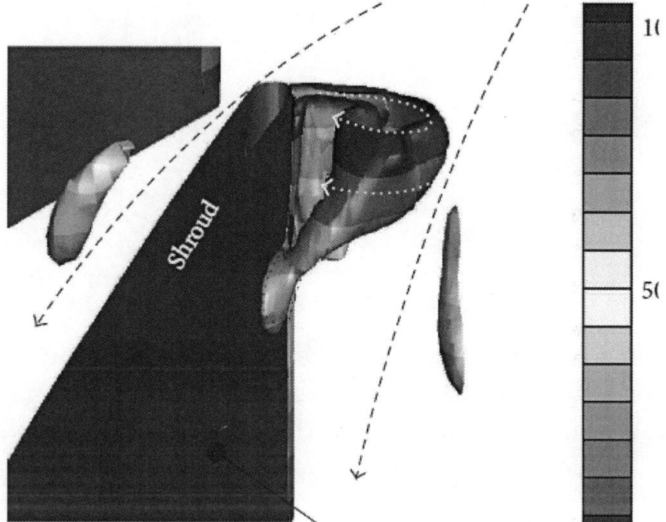

Figure 13. Instantaneous isosurface of the negative λ_2 vortex criterion.

Figure 13 shows the instantaneous negative isosurface of the second eigenvalue coloured by the normalized span. The separation due the high incidence angle allows the vorticity from the leading edge to be shed to form a vortex-tube. It spans from the diffuser vane suction side (60% span) to the shroud.Due to the large increase of incidence angle from80% span to the shroud, the upper end (near shroud) of the vortextube tends to move away from the diffuser vane suction side. The recent works of Pullan et al. [27] on axial configuration demonstrate that the process of spike formation is linked to a suction side boundary layer separation resulting from high incidence angle.

Figure 14. Instantaneous isosurface of the negative λ_2 vortex criterion representing the vortex formation.

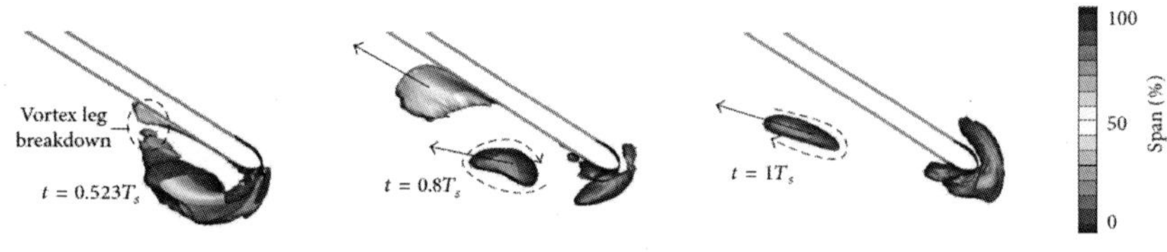

Figure 15: Instantaneous isosurface of the negative λ_2 vortex criterion representing the vortex displacement.

A vortex is also formed which starts at the suction surface and terminates at the shroud. It is observed that near the shroud the vortex moves in the circumferential direction along the shroud increasing the flow angle on the adjacent blade. The boundary layer on the adjacent blade separates and the structure can propagate according to the Emmons theory [5].The onset of spike instability has also been studied numerically by Everitt and Spakovszky [28] in an isolated vaned diffuser. The works show that flow separation at the diffuser vane leading edge associated to radial reverse flow near the shroud allows the vorticity from the leading edge to be convected into the vaneless space.The diffuser inlet blockage rises leading to diffuser instability. This study shows that the high tangential flow at the impeller exit is a key feature for spike onset. In the study configuration, due to the blade passing effects associated to the high flow angle fluctuations, the vortex core behavior is highly unsteady and time periodic with the blade passing frequency (a blade passage is composed of two channels). Therefore during one channel passing period the vortex forms and expands (Figure 14). During the second channel passing period (Figure 15), at $t = 0.0523T_S$ the vortex core is detached from the suction side, and the lower end separates from the upper end. As the radial velocity is higher at midspan compared to near the shroud, the lower end of the vortex is convected at higher velocity. It is found that the vortex-tube does not move in the circumferential direction but downstream along the suction side. Therefore, the scenario described for the spike onset in the literature is not observed for the simulated operating points. However, further mass flow reduction may lead to a more tangential flow (increase of flow angle) near the shroud inducing a vortex displacement toward the circumferential direction and initiating the spike inception process in the vaned diffuser.

CONCLUSION

Unsteady numerical simulations have been performed in a 2.5 pressure ratio centrifugal compressor stage to achieve a comprehensive description of the flow field from peak efficiency to near stall.Thecalculation domain extended to all the blade passages has permitted to observe a new unsteady flow pattern with a phenomenon decor related to the blade passing frequency when the compressor operates near stall. The conclusions are summarized as follows.

- In the impeller inducer, reducing the mass flow induces a rise of the secondary flow effects leading to a significant enlargement of the low momentum fluid region accumulated near the shroud. At near stall condition, the interface between the secondary and leakage flows with the main flow becomes unstable leading to a vortex formation.
- At the blade tip and at near stall conditions, the interface line that demarcates the oncoming flow from the leakage flow is almost aligned with the leading edge plane, suspecting that further mass flow reduction will drive the compressor into stall as observed in axial configurations.
- In the diffuser entry region, the decrease ofmass flow rate induces an increase of flow angle leading to a separation on the diffuser vane suction side near the shroud. At near stall condition, the vorticity from the separation is shed and leads

to a periodic formation of vortex-tube which travels along the suction side.

- Further mass flow reduction may change the vortex trajectory to a more tangential direction, leading to a propagation of the structure in the circumferential direction and inducing diffuser rotating stall. (v) Given theses conditions, the prediction of the stall onset region and a priori scenario appear difficult. However, recent investigations presume that the flow breakdown occurs in the impeller inducer due to the alignment of the interface line with the leading edge

NOMENCLATURE

Latin Letters

ent: Entropy (j/(kgK))

f, IF, BPF: Frequency (Hz), impeller frequency (Hz), and blade passing frequency (Hz)

m□ : Mass flow rate (kg/s)

MB, SB: Main blade, splitter blade

N: Number of blades

PS, SS: Pressure side, suction side

p: Pressure (Pa)

T: Time period (s), temperature (K)

t: Time(s)

NS: Near stall

OP: Operating point

U, V: Blade speed (m/s), absolute velocity (m/s)..

Superscripts and Subscripts

0: Total variable

1: Impeller inlet

2: Impeller exit

4: Diffuser exit

*: Reduced variable

b: Blade

m, r, t: Meridional, radial, and tangential

out: Domain outlet

R, S: Rotor,stator

Ref: Reference condition

f: Time-averaged value of f.

Greek Letters

α, β: Absolute flowangle (°), relative flow angle (°)

Ω: Rotational speed (rad/s).

CONFLICT OF INTERESTS

The authors declare that there is no conflict of interests regarding the publication of this paper.

ACKNOWLEDGMENT

The authors would like to express their thanks to Liebherr- Aerospace Toulouse SAS for supporting the present research program and to the CFD team of CERFACS for its help and support in the achievement of numerical simulations. The authors are also grateful to GENCI-CINES for providing computational resources.

REFERENCES

1. E. M. Greitzer, "Surge and rotating stall in axial flow compressors, part I: theoretical compression system model," ASME Journal of Engineering and Power, vol. 98, no. 2, pp. 190–198, 1976.
2. J. Galindo, J. R. Serrano, H. Climent, and A. Tiseira, "Experiments and modelling of surge in small centrifugal compressor for automotive engines," Experimental Thermal and Fluid Science, vol. 32, no. 3, pp. 818–826, 2008.
3.] N. Cumpsty, Compressor Aerodynamics, Pearson Education,2004.
4. G. J. Skoch, "Experimental investigation of centrifugal compressor stabilization techniques," Journal of Turbomachinery, vol. 125, no. 4, pp. 704–713, 2003.
5. H. W. Emmons, C. E. Pearson, and H. P. Grant, "Compressor surge

and stall propagation," Transaction of the ASME, vol. 77, pp. 455–469, 1955.

6. S. Mizuki and Y. Oosawa, "Unsteady flow within centrifugal compressor channels under rotating stall and surge," Journal of Turbomachinery, vol. 114, no. 2, pp. 312–320, 1992.
7.] M. P.Wernet, M. M. Bright, and G. J. Skoch, "An investigation of surge in a high-speed centrifugal compressor using digital PIV," Journal of Turbomachinery, vol. 123, no. 2, pp. 418–428, 2001.
8. I. Tr´ebinjac,N. Bulot, X.Ottavy, andN. Buffaz, "Surge inception in a transonic centrifugal compressor stage," in Proceedings of the ASME Turbo Expo, pp. GT2011-G45116, June 2011.
9. T. R. Camp and I. J. Day, "A study of spike and modal stall phenomena in a low-speed axial compressor," Journal of Turbomachinery, vol. 120, no. 3, pp. 393–401, 1998.
10. Z. S. Spakovszky and C. H. Roduner, "Spike and modal stall inception in an advanced turbocharger centrifugal compressor," Journal of Turbomachinery, vol. 131, no. 3, pp. 1–9, 2009.
11. L. Cambier and M. Gazaix, "elsA: an efficient object-oriented solution to CFD complexity," in Proceedings of the 40th Aerospace Science Meeting and Exhibit, Reno, Nev, USA, 2002.
12. P. R. Spalart and S. R. Allmaras, "One-equation turbulence model for aerodynamic flows," Recherche Aerospatiale, no. 1, pp. 5–21, 1994.
13. S. Yoon and A. Jameson, "An LU-SSOR scheme for the Euler and Navier-Stokes equation," in Proceedings of the AIAA 25th Aerospace Science Meeting, Paper No. 87-0600, Reno, Nev, USA, 2002.
14. A. Jameson, "Time dependent calculations using multigrid, with applications to unsteady flows airfoils and wings," in Proceedings of the 10th AIAA Computational Fluid Dynamics Conference, Paper No. 91-1596, Reno, Nev, USA, 1991.
15. F. Sicot,G.Dufour, andN.Gourdain, "Atime-domain harmonic balance method for rotor/stator interactions," Journal of Turbomachinery, vol. 134, no. 1, Article ID 011001, 13 pages, 2012.
16. G. Filola, M. C. Le Pape, and M. Montagnac, "Numerical simulations around wing control surfaces," in Proceedings of the International Conference on Agricultural Statistics (ICAS '04), 2004.
17. N. Gourdain, M. Montagnac, F. Wlassow, and M. Gazaix, "High-performance computing to simulate large-scale industrial flows in multistage compressors," International Journal of High Performance Computing Applications, vol. 24, no. 4, pp. 429–443, 2010.
18. G. Dufour, X. Carbonneau, P. Arbez, J. Cazalbou, and P. Chassaing,

"Mesh-generation parameters influence on centrifugal compressor simulation for design optimization," in Proceedings of the ASME Heat Transfer/Fluids Engineering Summer Conference (HT/FED '04), Paper No. 8004-56314, pp. 609–617, July 2004.

19. D. Eckardt, "Instantaneous measurements in the jet wake discharge flow of a centrifugal compressor impeller," ASME Journal of Engineering for Power, vol. 97, pp. 337–346, 1975.
20. D. Eckardt, "Detailed flow investigations within a high-speed centrifugal compressor impeller," Journal of Fluids Engineering, vol. 98, no. 3, pp. 390–402, 1976.
21. W. R. Hawthorne, "Secondary vorticity in stratified compressible fluids in rotating systems," CUED/A-Turbo/TR 63, University of Cambridge, Cambridge, UK, 1974.
22. J.M¨arz, C. Hah, andW. Neise, "An experimental and numerical investigation into the mechanisms of rotating instability," Journal of Turbomachinery, vol. 124, no. 3, pp. 367–374, 2002.
23. H. Duc Vo, C. S. Tan, and E. M. Greitzer, "Criteria for spike initiated rotating stall," Journal of Turbomachinery, vol. 130, no. 1, Article ID 011023, 5 pages, 2008.
24. R. C. Dean and Y. Senoo, "Rotating wakes in vaneless diffusers," Journal of Basic Engineering, vol. 82, pp. 573–574, 1960.
25. S. Deniz, E. M. Greitzer, and N. A. Cumpsty, "Effects of inlet flow field conditions on the performance of centrifugal compressor diffusers, part 2: straight-channel diffuser," Journal of Turbomachinery, vol. 122, no. 1, pp. 11–21, 2000.
26. J. J. Jinhee Jeong and F. Hussain, "On the identification of a vortex," Journal of Fluid Mechanics, vol. 285, pp. 69–94, 1995.
27. G. Pullan, A. M. Young, I. J. Day, E. M. Greitzer, and Z. S. Spakovszky, "Origins and structure of spike-type rotating stall," in Proceedings of the ASME Turbo Expo, pp. GT2012-G68707, June 2012.
28. J. N. Everitt and Z. S. Spakovszky, "An investigation of stall inception in centrifugal ompressor vaned diffusers," in Proceedings of the ASME Turbo Expo, pp. GT2011-G46332, June 2011.

Chapter 7

EFFECT OF MECHANISM ERROR ON INPUT TORQUE OF SCROLL COMPRESSOR

Man Zhao, Shurong Yu, Chao Li, and Yang Yu

College of Petrochemical Technology, Lanzhou University of Technology, Lanzhou 730050, China

ABSTRACT

Based on the fundamental principle of plane four-bar mechanism, the force on the equivalent parallel four-bar mechanism was analyzed for scroll compressor with mini-crank ant rotation, and the formula of input torque was proposed. The change of input torque caused by the mechanism size error was analyzed and verified with an example. The calculation results show that the mechanism size error will cause large fluctuation in input torque at the drive rod and connecting rod collinear and the fluctuation extreme value increases with rotational speed. Decreasing of the crankshaft eccentricity errors is helpful for reducing the effects of dimension error on input torque but will increase the friction loss of orbiting and fixed scroll wrap. The influence of size error should be considered in design in order to select suitable machining accuracy and reduce the adverse effect caused by size error.

INTRODUCTION

With the needs of clean and oil-free pollution compressed air in food, medicine, fuel cells, and other industries, the investigation of oil-free scroll compressor research has been one of hot topics in scroll compressors. The mini-crank ant rotation mechanism is often used to achieve the orbiting scroll work under the oil-free lubrication conditions. The movement relationship of mini-crank, orbiting scroll, crankshaft, and the bracket body can be represented by planar four-bar mechanism [1], and the ideal working should be planar parallel four-bar mechanism. But in fact, machining error and running wear are inevitable, such as the orbiting scroll deviating from the ideal state, resulting in excessive friction wear or large tangential leaking [2], also causing the movable linkages uneven movement. The research of the size error on the four-bar mechanism critical rod's uneven movement and the change of input torque can provide some theoretical foundation for machining accuracy and scroll compressor reliable operation.

FOUR-BAR LINKAGE MECHANISM

The structure of scroll compressor with mini-crank anti rotation is shown in Figure 1. The mini-crank is respectively connected with the orbiting scroll and the bracket body through the bearing. In order to ensure the orbiting scroll's revolving translational motion, the gyration radius of mini-crank and eccentric crankshaft should be in the same size; that is, the plane equivalent four-bar mechanism of scroll compressor, shown in Figure 2, should be the parallel four-bar mechanism

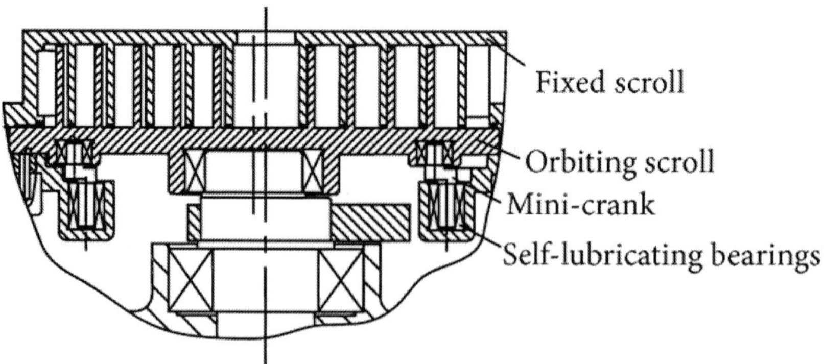

Figure 1: Configuration diagram of scroll compressor.

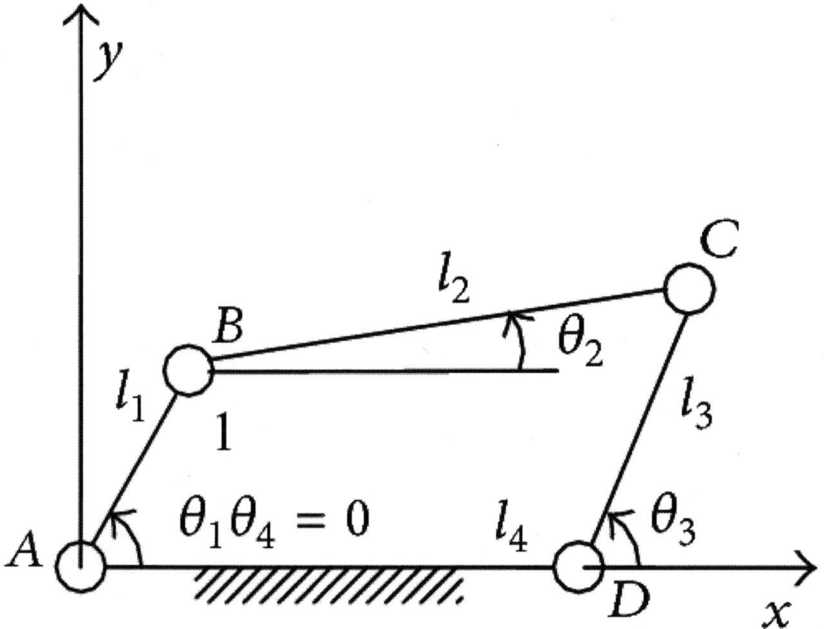

Figure 2: Four-bar mechanism.

Figure 2 illustrates the four-bar mechanism in the scroll compressor work,l1 is equivalent to the drive crankshaft, is equivalent to the orbiting scroll, is equivalent to the mini-crank, and is l4 equivalent to the bracket body. In ideal working conditions: l1=l3,l2=l4

Geometrical Theory

Cartesian coordinate system is shown in Figure 2; the vector equation is

$$\begin{aligned} l_1 \cos\theta_1 + l_2 \cos\theta_2 &= l_3 \cos\theta_3 + l_4, \\ l_1 \sin\theta_1 + l_2 \sin\theta_2 &= l_3 \sin\theta_3. \end{aligned} \tag{1}$$

Equation (1) can be written as

$$A\cos\theta_2 + B\sin\theta_2 = C, \quad (2)$$

$$A\cos\theta_3 + B\sin\theta_3 = D, \quad (3)$$

Where

$$\begin{aligned} A &= l_4 - l_1\cos\theta_1, \\ B &= -l_1\sin\theta_1, \\ C &= \frac{1}{2l_2}\left(l_2^2 - l_3^2 + A^2 + B^2\right), \\ D &= \frac{1}{2l_3}\left(l_2^2 - l_3^2 - A^2 - B^2\right). \end{aligned} \quad (4)$$

According to (2) and (3), the output angles ø2 and ø3 of four-bar mechanism can be expressed as

$$\begin{aligned} \theta_2 &= 2\arctan\frac{B - M\sqrt{A^2 + B^2 - C^2}}{A + C}, \\ \theta_3 &= 2\arctan\frac{B - M\sqrt{A^2 + B^2 - D^2}}{A + D}. \end{aligned} \quad (5)$$

The output angles ø2 and ø3 of the mechanism have two solutions, and the exact value can be determined according to the mechanism's initial installation and the continuity of the movement. When the hinge point B, C, and D are arranged clockwise M will take the positive sign; when the hinge points B ,C , and are arranged anticlockwise, has a negative sign.

Taking first derivative and second derivative of vector equation (1), the angular velocity and angular acceleration of rod 2 and rod 3 can be obtained:

$$\dot{\theta}_2 = \frac{l_1 \sin(\theta_1 - \theta_3)}{l_2 \sin(\theta_3 - \theta_2)}\dot{\theta}_1,$$

$$\dot{\theta}_3 = \frac{l_1 \sin(\theta_1 - \theta_2)}{l_3 \sin(\theta_3 - \theta_2)}\dot{\theta}_1,$$

$$\ddot{\theta}_2 = \frac{-\dot{\theta}_1^2 l_1 \cos(\theta_1 - \theta_3) - \dot{\theta}_2^2 l_2 \cos(\theta_2 - \theta_3) + \dot{\theta}_3^2 l_3}{l_2 \sin(\theta_2 - \theta_3)}, \tag{2.1}$$

$$\ddot{\theta}_3 = \frac{\dot{\theta}_1^2 l_1 \cos(\theta_1 - \theta_2) + \dot{\theta}_2^2 l_2 - \dot{\theta}_3^2 l_3 \cos(\theta_3 - \theta_2)}{l_3 \sin(\theta_3 - \theta_2)}.$$

For the parallel four-bar mechanism, when the drive rod is running at a constant angular velocity, the following results can be obtained:

$$\dot{\theta}_2 = 0, \quad \ddot{\theta}_2 = 0, \quad \dot{\theta}_3 = \dot{\theta}_1, \quad \ddot{\theta}_3 = 0. \tag{7}$$

Mechanism Motion Analysis

By the scroll compressor structure, the active bars centroid coordinates are

$$\begin{aligned} x_{S2} &= x_{S1} = l_1 \cos\theta_1, \\ y_{S2} &= y_{S1} = l_1 \sin\theta_1, \end{aligned} \tag{8}$$

$$\begin{aligned} x_{S3} &= l_4 + l_3 \cos\theta_3, \\ y_{S3} &= l_3 \sin\theta_3. \end{aligned} \tag{9}$$

Respectively, by seeking the first derivative and second derivative of (8) and (9), the angular velocity and angular acceleration of the movable rods can be obtained

MECHANICAL MODELS

While four-bar mechanism is working, in addition to the input torque, the force, and moment acting on mechanism are mainly three types: each movable rod gravity, inertia force, inertia moment; external forces and moment; hinge friction and friction moment.

For the equivalent four-bar mechanism of scroll compressor, external loading in radial plane is shown in Figure 3; tangential gas force , radial gas force , rotation torque M_z, F_{fd1}, F_{fd2}, the friction force of the orbiting plate and fixed plate, as well as the gravity, inertia force and inertia moment used to balance the rotor system [3].

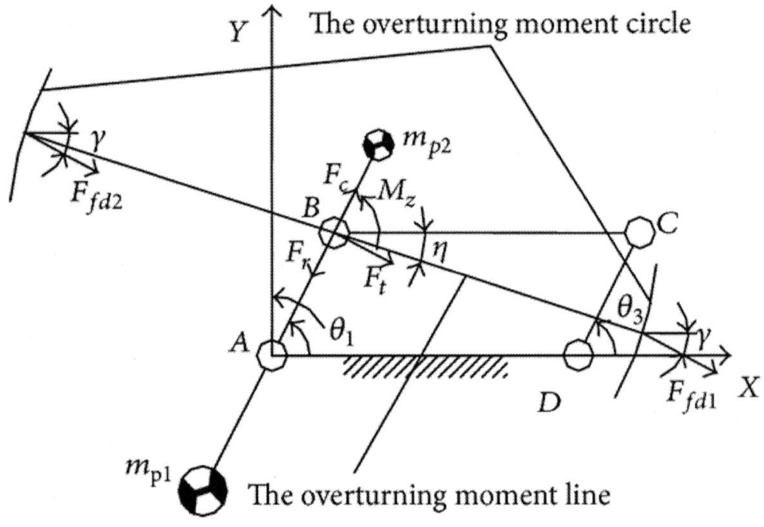

Figure 3: Schematic diagram of external loading.

The geometric parameters of balance weights Mp1 and Mp2 are determined by the rotor system balance. The relationship between the mass of the crankshaft pin, the orbiting scroll assembly, and the balance weights is

$$(m_1 + m_2) l_1 + m_{p2} r_{p2} = m_{p1} r_{p1}, \qquad (10)$$

Where Rp1 is the distance between the centroid of counterweight Mp1and Mp2 the rotation center Mp2 is the distance between the centroid of counterweight and the rotation center.

The orbiting scroll revolving translational motion characteristics shows that any point of the orbiting scroll is in the same movement rule. The direction of the frictional force is opposite to the orbiting scroll gravity (hinge B) moving, and the force role point is on the overturning moment action line [4]:

$$\begin{aligned} F_{fd1} &= \mu_d F_{d1}, \\ F_{fd_2} &= \mu_d F_{d2}. \end{aligned} \tag{11}$$

Msd, the friction torque on the orbiting scroll center (hinge B) caused by the friction forces $Ffd1$ and $Ffd2$, is as follows:

$$M_{sd} = \mu_d R_t (F_{d1} - F_{d2}) \sin(\eta - \gamma), \tag{12}$$

Where η is the azimuth angle between the action line and the x-axis; γ is the azimuth angle between the direction of $F\,fd$ and the x-axis; R t is the radius of overturning moment action circle; $\mu\ d$ is the friction coefficient of orbiting scroll and fixed scroll. F

$d1$ and $F\ d2$ can be obtained by the axial force balance of the orbiting scroll [5]:

$$F_{d1} - F_{d2} = -\frac{M_m}{R_t}. \tag{13}$$

So, (12) can be rewritten as

$$M_{sd} = \mu_d M_m \sin(\gamma - \eta), \tag{14}$$

Where *M m* is the overturning moment acting on the bearing drive surface.

INPUT TORQUE

Excluding the influence of friction hinge deputy, the input torque, *T*in, can be considered to have three parts:

- *T d*, inertia torque, to overcome inertia influence to maintain the driving rod uniform speed;
- *T w*, gravitational torque, to balance the effect of gravity;
- *T F*, external load torque, to balance the outside load.

For the scroll compressor, with uniform speed and conducted rotor system balance design, by the virtual work equation [6], input torque is as follows:

$$
\begin{aligned}
T_d &= -\frac{J_2\ddot{\theta}_2\delta\theta_2}{\delta\theta_1} + \frac{J_3\ddot{\theta}_3\delta\theta_3}{\delta\theta_1},\\
T_w &= \frac{m_3 g l_3 \cos\theta_3 \delta\theta_3}{\delta\theta_1},\\
T_F &= M_t - \frac{(M_z + M_{sd})\,\delta\theta_2}{\delta\theta_1},
\end{aligned}
\tag{15}
$$

whereMt= $F(\theta1)l1$, which is the resistance moment caused by the tangential gas force acting on the crankshaft rotation center through the orbiting scroll. $\delta\theta 2$ and $\delta\theta 3$ are given by the formulae (2) and (3) variation solution, respectively:

$$
\begin{aligned}
\delta\theta_2 &= \frac{l_1[l_2\sin(\theta_2-\theta_1)+l_4\sin\theta_1]}{l_2[l_1\sin(\theta_2-\theta_1)-l_4\sin\theta_2]}\delta\theta_1,\\
\delta\theta_3 &= \frac{l_1[l_3\sin(\theta_3-\theta_1)-l_4\sin\theta_1]}{l_3[l_1\sin(\theta_3-\theta_1)-l_4\sin\theta_3]}\delta\theta_1.
\end{aligned}
\tag{16}
$$

The mini-crank quality ($m3$) could be negligible in the dynamical analysis, due to the structural characteristics of the scroll compressor.

Thus, when the scroll compressor is in a uniform speed, the input torque can be expressed as

$$T_{in}(\theta_1) = M_t - \frac{\left[M_z + J_2\ddot{\theta}_2 + M_{sd}\right]\delta\theta_2}{\delta\theta_1}. \quad (17)$$

When a scroll compressor is in ideal working condition, □2= 0, $\delta\theta 2$= 0, the input torque is

$$T_{in}(\theta_1) = M_t. \quad (18)$$

Kinematics and kinetics show that the conditions of drive rod doing whole rotary motion [7] are(1)the sum of the shortest and longest rod length should be less than or equal to the length of the remaining two;(2)the shortest rod is connected frame bar.

The scroll compressor equivalent four-bar mechanism shows that regardless of the existence of rod size error, *l*1and *l*4, rotation consisting deputy vice must be a complete rotation.

So,

$$l_1 + l_2 \le l_3 + l_4 \quad (19)$$

Or

$$l_1 + l_4 \le l_3 + l_2. \quad (20)$$

And when the actual length of *l*3 is less than *l*1, there must be clearances that exist in hinged deputy to ensure the drive rod doing whole rotary motion

MODEL VERIFICATION

Taking a twin-wraps scroll compressor as an example, we analyzed the effect of size error on input torque, without considering the mechanism clearances and hinge friction. Only take the shortest lever $l1$ into account. Table 1 shows the parameters of the equivalent four bar mechanism parameters for input torque analysis. The external load changing with the crank angle of the examples of scroll compressor is shown in Figures 4 and 5; where radial gas force $F\ r$ and orbiting scroll assembly inertial force Fc are certain values at a certain speed, and the inertial force $F\ c$ increases rapidly with the speed. The rotation moment $M\ Z$ and the resistance moment $M\ t$ change with the crank angle θ at the same discipline of tangential gas force $F\ t$. The friction moment Msd increases with the speed quickly but is approximately constant at a certain speed. It can be seen from Figures 6, 7, and 8:

- As the scroll compression is in ideal working condition, the variation of input torque of scroll compression is the same with the resistance moment $M\ t$, as shown in Figure 6 with slight variation. So, the scroll compressor is designed generally without flywheel setting needed;
- When the equivalent parallel four linkage mechanism size error occurs, the actual input torque of compressor differs from the ideal condition a lot.

Table 1: Parameters of the equivalent parallel four-bar mechanism.

Parameters	Values
l_1 (mm)	4.5
l_2 (mm)	115
l_3 (mm)	4.5
l_4 (mm)	115
J_2 (Kg·m^2)	0.142

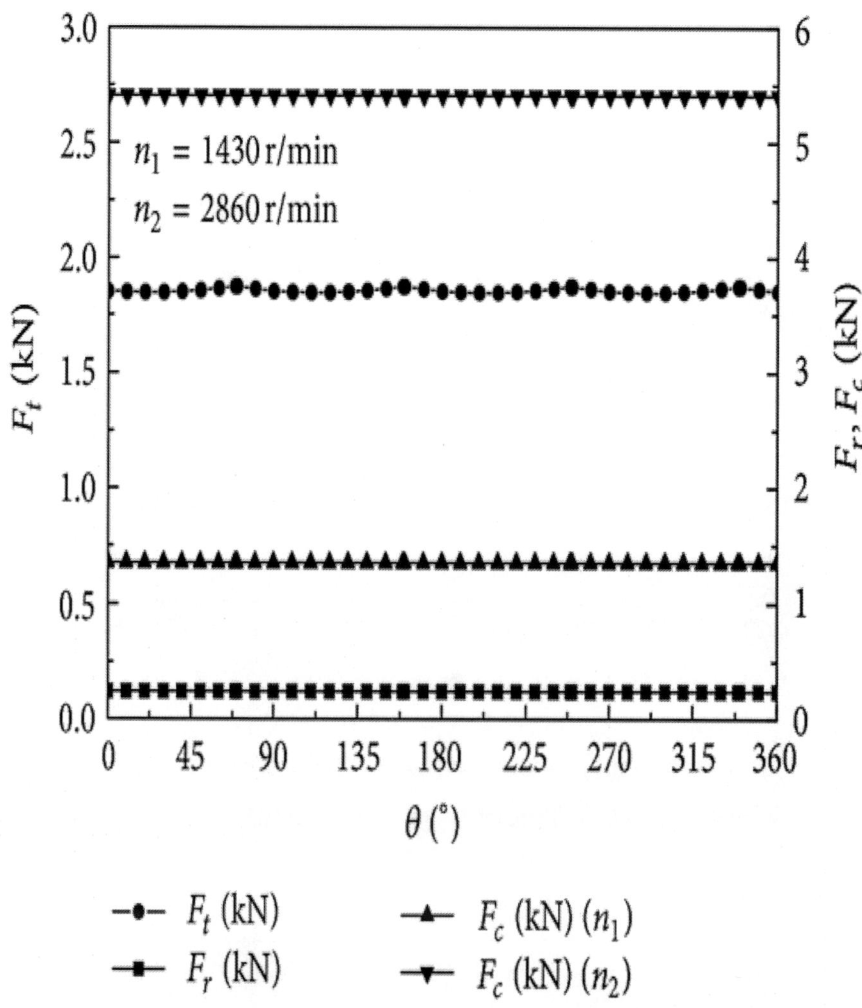

Figure 4: The diagram of external force.

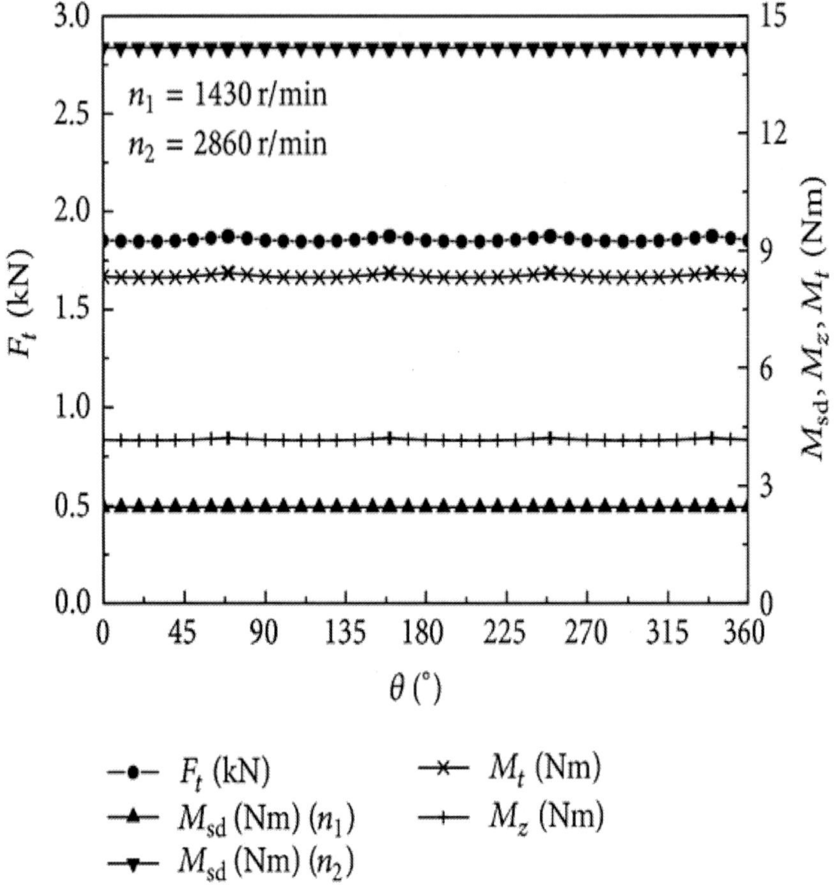

Figure 5: The diagram of external moment.

It can be seen from Figures 6, 7, and 8:(1)as the scroll compression is in ideal working condition, the variation of input torque of scroll compression is the same with the resistance moment , as shown in Figure 6 with slight variation. So, the scroll compressor is designed generally without flywheel setting needed;(2)when the equivalent parallel four linkage mechanism size error occurs, the actual input torque of compressor differs from the ideal condition a lot.

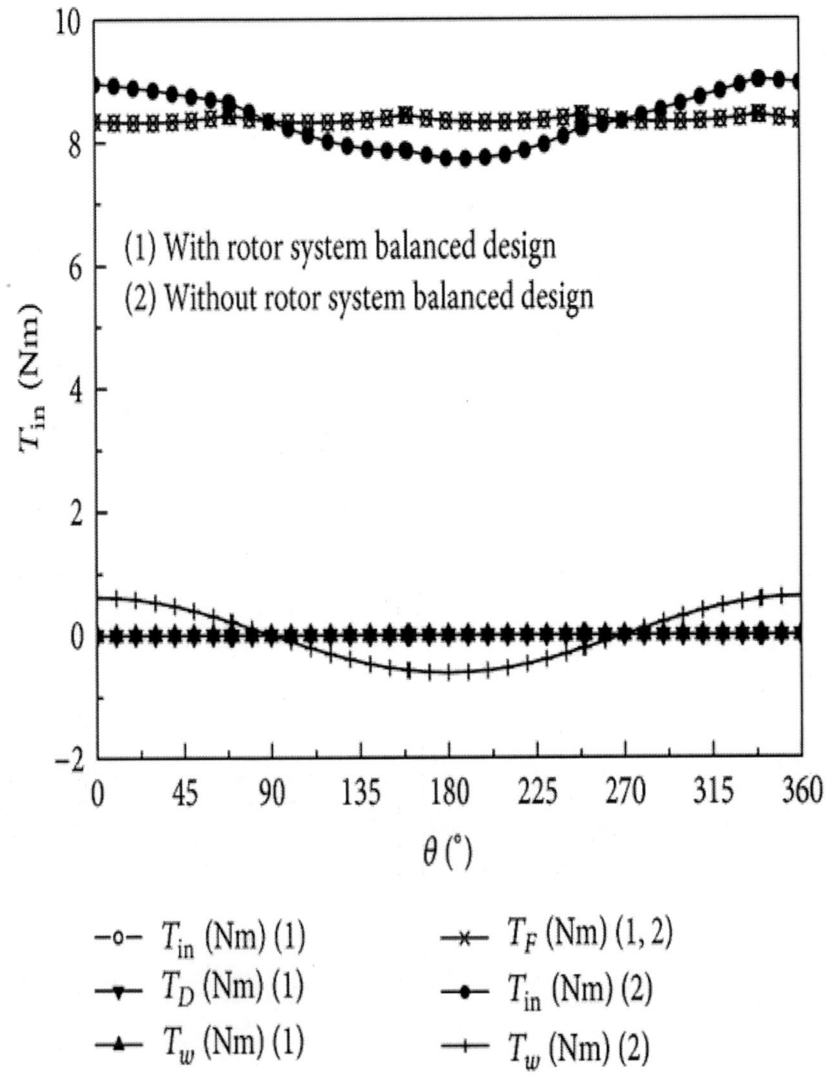

Figure 6: The diagram of input torque in ideal working.

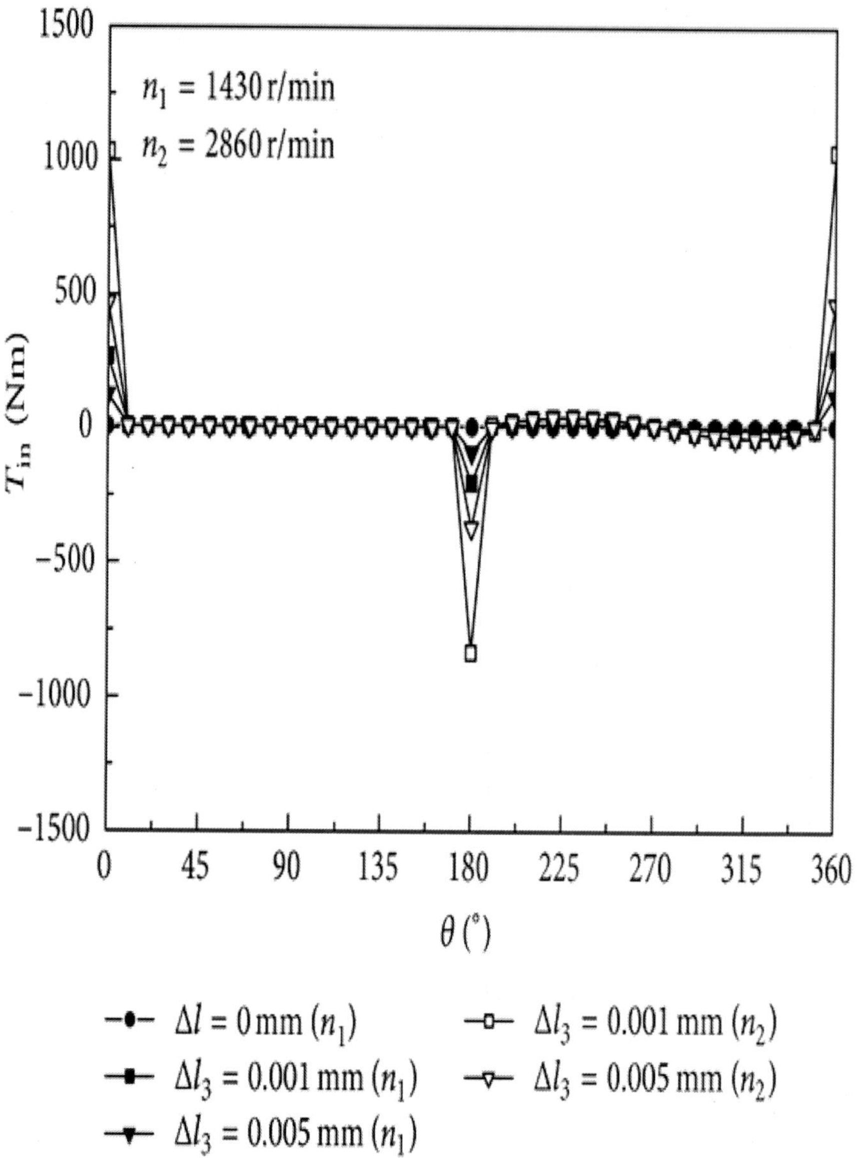

Figure 7: The diagram of the input torque

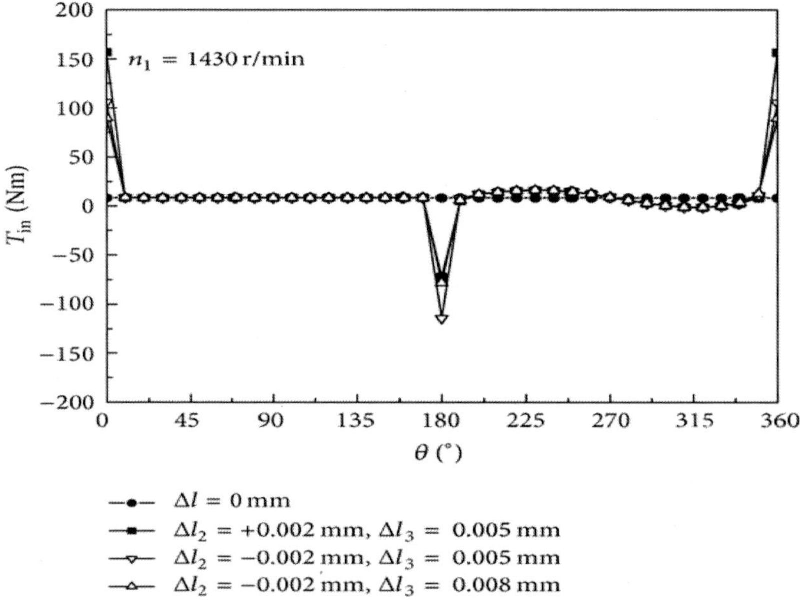

Figure 8: The comparison of input torque.

On the size error and crank shaft speed. But, under different conditions, the value is insignificant and little change occurs in the amplitude in the region.

Second, the input torque will cause great fluctuation as mini-crank rotation direction changes. Comparison shows that the mini-crank changes in the same direction as the crankshaft, extreme a fluctuation is the maximum and the higher the speed, the greater extremes. Meanwhile, the scroll compressor structure determines that the more positive deviation *l*2 (or negative deviation *l*4),the more extreme fluctuations increase and the more negative deviation *l*1 (or positive deviation *l*3), the more extreme Fluctuations reduce.

CONCLUSION

Based on the fundamental principle of plane linkage mechanism, effect of mechanism error on input torque of a completely dynamic balancing of scroll compressor was analyzed. Mechanism size error has a big influence on the drive rod and connecting rod collinear. The drive crankshaft eccentricity is beneficial for error reduction but causes more friction loss in the wrap of orbiting and fixed scroll. The appropriate precision should be selected, reducing the adverse effect caused by size error.

CONFLICT OF INTERESTS

The authors declare that there is no conflict of interests regarding the publication of this paper.

ACKNOWLEDGMENT

This work was supported by the National Natural Science Foundation of China (no. 51265026).

REFERENCES

1. Z. Liu, J. Wang, and J. Qiang, Scroll Fluid Machinery and Scroll Compressor, Machinery Industry Press, Beijing, China, 2009.
2. C. Li, R. Zhao, and Z. Liu, "Effect of mechanism error of scroll compressor on sealing clearance," Lubrication Engineering, vol. 32, no. 7, pp. 66–68, 71, 2007.
3. M. Zhao, S. Yu, C. Li, and Y. Yu, "The rotor system balance design of scroll fluid machinery," Applied Mechanics and Materials, vol. 160, pp. 268–272, 2012. View at Publisher · View at Google Scholar · View at Scopus
4. M. Zhao, S. Yu, C. Li, and Y. Yu, "Analysis of orbiting scroll overturning for scroll compressor," Applied Mechanics and Materials, vol. 226–228, pp. 576–579, 2012.
5. Z. Gu, Y. Yu, and S. Feng, Scroll Compressor and Other Scroll Machinery, Shanxi Science and Technology Press, Xi'an, China, 1998.
6. J. Sun and C. Zhang, Structural Mechanics, Chongqing University Press, 2001.
7. H. Sun, Mechanical Principles, Higher Education Press, Beijing, China, 2001.

Chapter 8

OPTIMUM DESIGN OF OIL LUBRICATED THRUST BEARING FOR HARD DISK DRIVE WITH HIGH SPEED SPINDLE MOTOR

Yuta Sunami,[1] Mohd Danial Ibrahim,[2] and Hiromu Hashimoto[1]

[1]Department of Mechanical Engineering, Tokai University, 4-1-1 Kitakaname, Hiratsuka-shi, Kanagawa-ken 259-1292, Japan

[2]Department of Mechanical and Manufacturing Engineering, Faculty of Engineering, University Malaysia Sarawak, 94300 Kota Samara Han, Sarawak, Malaysia

ABSTRACT

This paper presents the application of optimization method developed by Hashimoto to design oil lubricated thrust bearings for 2.5 inch form factor hard disk drives (HDD). The designing involves optimization of groove geometry and dimensions. Calculations are carried out to maximize the dynamic stiffness of the thrust bearing spindle motor. Static and dynamic characteristics of the modeled thrust bearing are calculated using the divergence formulation method. Results show that, by using the proposed optimization method, dynamic stiffness values can be well improved with the bearing geometries not being fixed to conventional grooves.

INTRODUCTION

HDD has been used as the main storage multimedia for electronics devices. Currently, HDD widely depends on oil lubricated hydrodynamic bearings. Hydrodynamic bearing is mainly supported by thrust and journal bearings. A schematic view of bearings in a 2.5 inch HDD is shown in Figures 1 and 2.

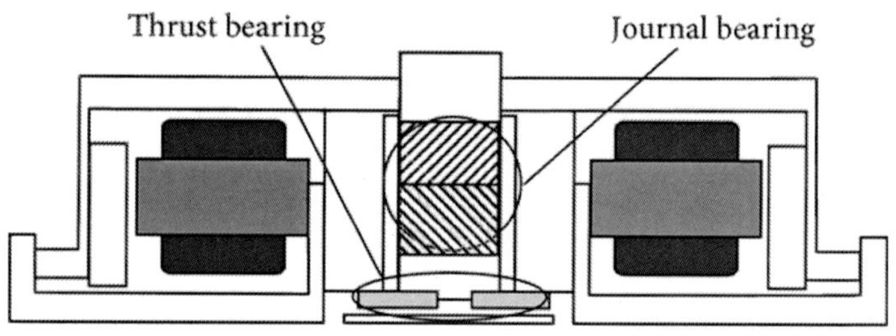

Figure 1. Bearings in spindle motor.

(a)

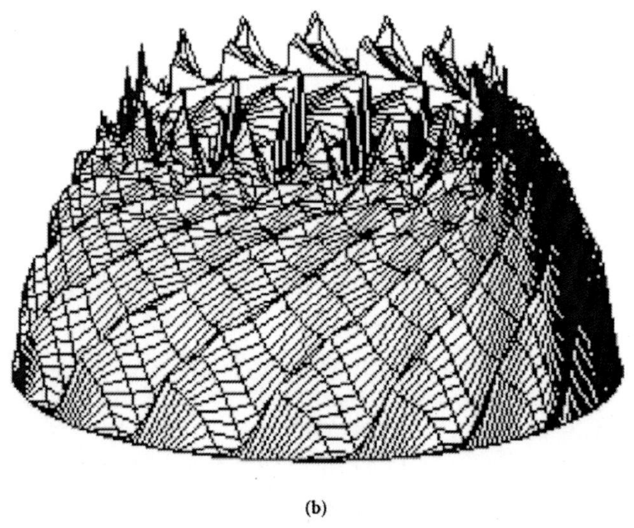

(b)

Figure 2. Initial spiral thrust bearing groove geometry and pressure distribution; (a) groove geometry, (b) pressure distribution.

Hydrodynamic bearing gives better performance characteristics compared to conventional ball bearings as it has high dynamic stiffness with additional much higher damping effects. These damping effect characteristics provide smaller no repeatable run-out (NRRO). NRRO is the major contributor to the track miss registration in HDD read-write mechanism. Even though the repeatable run-out (RRO) of oil lubricated hydrodynamic bearing is higher than the ball bearing spindle motor, the RRO can be corrected by a read-write servo. Therefore, to reduce NRRO and increase spindle performance is to improve the bearing performance.

Currently, groove geometries that are being widely used in HDD thrust bearings are mainly a spiral or a herringbone grooved geometries. Several investigators have conducted numerical analysis predictions of these grooves for HDD bearing performance [1–5]. There were also some attempts at improving the dynamic stiffness and damping of HDD spindle by introducing permanent magnetic thrust plates into the bearing spindle structure [6] or introducing magnetic fluid as lubricants [7]. However, there are very few attempts at finding an optimum design with a novel geometry to replace the conventional herringbone or spiral grooves for HDD.

In the attempts at improving HDD performance, Arakawa et al. [8] proposed a no uniform spiral groove for journal bearings to expand the critical bearing number for higher revolution speed of HDD spindle. The approach indicated that the novel no uniform spiral grooves hydrodynamic bearings manage to increase the stability of rotation in high speeds.

This suggests that there is a probability of a further improvement if the groove geometry is not being fixed to any conventional grooves, either spiral or herringbone grooves. However, as far as authors know, there are no attempts at drastically improving the bearing characteristics changing the geometry and dimensions for oil lubricated 2.5 inch HDD spindle motor. Recently, HDD is being demanded to be thinner. If the characteristic of thrust bearing can be drastically improved, it is possible for HDD to be thinner. Therefore, in this paper, by adapting the optimization method initiated by Hashimoto, new optimum groove geometry and dimension to replace conventional grooves and increase the bearing performance of oil lubricated 2.5 inch HDD had been calculated. The groove geometry and dimension of the thrust bearing for spindle motor are calculated using the hybrid method [9, 10] with the improvement of dynamic stiffness, k being set as the objective function.

The optimization results showed that the dynamic stiffness of oil lubricated thrust bearing can be improved by introducing a new optimum geometry into the system. Vibration analyses were also presented to numerically verify the applicability of the improved bearings.

GEOMETRICAL OPTIMIZATION AND ANALYSIS OF BEARING CHARACTERISTICS

Modification of Groove Geometry

The geometrical optimization process is carried out using the hybrid method [9, 10], a combination of direct search method and successive quadratic programming (SQP) method. SQP is a powerful method to obtain the optimum solutions for nonlinear optimization problems. However, it sometimes cannot find the global optimum solution when the objective function has the multipacks. Therefore, the direct search method is used first to roughly search for several local optimum solutions; then, SQP is used to find the local optimum solutions when we set the initial values obtained by the direct search method. The global optimum solution is finally obtained by choosing the maximum or minimum value. Furthermore, the confirmation of the optimum solution is needed to clarify the validity of the optimum solution. The validity of the optimum solution is confirmed by comparing the value of the objective function calculated from the optimum design variables with those from randomly selected design variables.

Figure 3 shows the design parameters used in this optimization. The initial prescribed design parameters for thrust bearing in this paper, based on a commercialized 2.5 inch form factor HDD design, are shown in Table 1. Taking the spiral groove geometry as threshold groove geometry, the groove geometry parameters of angle variation $\varphi 3$ through $\varphi 4$ are initially set to zero.

Table1. Initial 2.5 inch bearing parameters

Parameters	Values
Outer radius R_{out} (m)	2.55×10^{-3}
Inner radius R_{in} (m)	1.0×10^{-3}
Weight W_d (N)	0.185
Revolution speed n_s (rpm)	5000~20,000
Groove number N	16
Seal ratio r_s	0.58
Groove depth h_g (m)	1.0×10^{-5}
Groove width ratio $\alpha = (b_1/b_1 + b_2)$	0.5
Geometry parameter φ_1 (deg)	0.0
Geometry parameter φ_2 (deg)	0.0
Geometry parameter φ_3 (deg)	0.0
Geometry parameter φ_4 (deg)	0.0
Lubricant viscosity μ (Pa·s)	6.523×10^{-3}

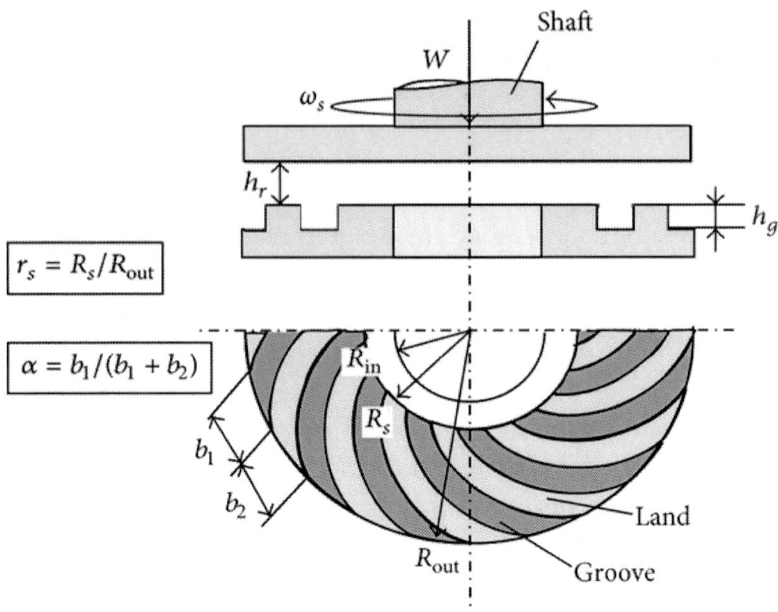

Figure 3. Design parameters for thrust bearing.

From there, the groove geometry design parameters are changed arbitrarily using the cubic spline interpolation function [9]. The cubic spline interpolation function is a cubic polynomial equation where each section connected to nodal points is a continuity of a second order derivative of the function.

Figure 4 shows the spiral groove geometry represented as initial values in the feasible region for optimum design of thrust bearing treated in this paper. Referring to the same figure, the initial line representing the spiral groove geometry, is shown with the label initial groove geometry while the new line of geometry after the spline interpolation function is calculated step by step as shown with the label th step. The groove geometry parameter of $\varphi 1$through $\varphi 2$ is set starting from inner region towards the outer region. The feasible region for the groove geometry is divided equally into four regions in the radial direction as there were no significant effects on the optimization with regions over the divisions of four.

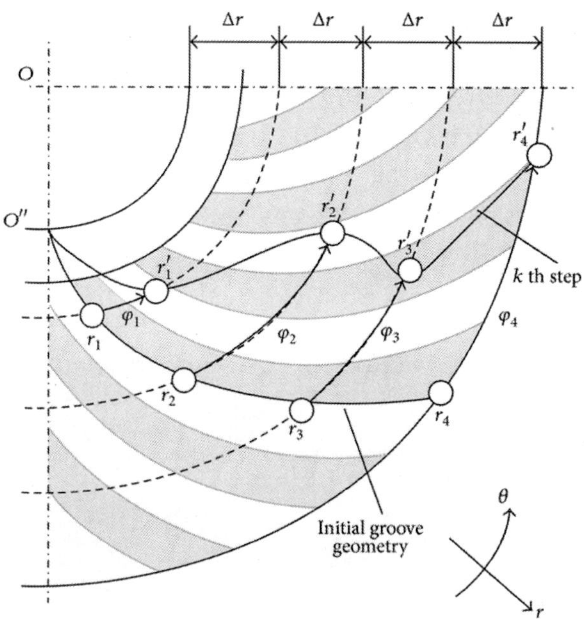

Figure 4. Modifications of groove geometry.

Analysis of Bearing Characteristics

In this paper, applying the boundary-fitted coordinate system to adjust the geometrical optimization method derives the calculation method of bearing characteristics. Moreover, in the process of analyzing the static and dynamic characteristics of the bearings, the perturbation method is applied to the Reynolds equivalent equation. The Reynolds equivalent equation can be solved using the Newton-Rap son iteration method, and the static component *p*0 and dynamic component *pt* of pressure are obtained. This detailed calculation method is shown in the appendix.

The load-carrying capacity*W*is obtained by the following integration:

$$W = \int_0^{2\pi} \int_{r_1}^{r_2} \{p_0(h_{r0}) - p_a\}\, r dr\, d\theta, \tag{1}$$

Where pa indicates the atmospheric pressure.

The minimum oil lubricating film thickness hr is simultaneously determined from the equilibrium condition between the axial load acting on a thrust bearing and the bearing load carrying capacity. The minimum oil lubricating film thickness hr is obtained by solving the following force balance equation:

$$W(h_r) - mg = 0. \quad (2)$$

The spindle motor's friction torque on bearing surface is given by the following integration:

$$T_r = \int_{r_1}^{r_2} \int_0^{2\pi} \left(\frac{\mu r^3 \omega_s}{h_0} - \frac{r h_0}{2} \frac{\partial p_0}{\partial \theta} \right) dr\, d\theta, \quad (3)$$

Where ωs is the angular velocity of the rotating shaft

The spring coefficient k and damping coefficient c can be obtained, respectively, by integrating the real and imaginary parts of the dynamic pressure components, pt, as follows:

$$k = \int_{r_1}^{r_2} \int_0^{2\pi} \mathrm{Re}(-p_t)\, r d\theta\, dr$$
$$c = \frac{1}{\omega_f} \int_{r_1}^{r_2} \int_0^{2\pi} \mathrm{Im}(-p_t)\, r d\theta\, dr, \quad (4)$$

where ωf is the squeeze frequency of the shaft, which is angular frequency of harmonic vibration when the bearing is small vibrated in the axial direction under equilibrium condition, in the HDD obtained through the perturbation method the and modified Reynolds equation [9].

Finally, the bearing stiffness, K, is given by the following equation

$$K = \sqrt{k^2 + (\omega_f \cdot c)^2}. \quad (5)$$

In this paper, the oil leak is not considered because the surround of thrust bearing of the spindle motor for 2.5 inch HDD, as shown in Figure 1, is sealed.

FORMULATION OF OPTIMUM DESIGN AND VIBRATION ANALYSIS

Formulation of Optimum Design

In formulating the optimum design problems for groove geometry, the design variable of vector X consisting of groove geometry design variables of angle variations $\varphi 1$ through $\varphi 4$ and groove dimension design variables of *rs, hg,* and α as shown in Figures 3 and 4 is defined as follows:

$$\mathbf{X} = \left(r_s, h_g, \alpha, \varphi_1, \varphi_2, \varphi_3, \varphi_4\right), \tag{6}$$

where *hg*, *rs*= (*Rs*/*R*out), and α = (*b*1/(*b*1+ *b*2)) are the groove depth, the seal ratio, and the ratio of groove width, respectively. The seal ratio means ratio of seal part, and the ratio of groove width means ratio of groove part of the total surface area.

Another design variable considered in this optimization is the number of grooves. However, since design variables of groove numbers are of discrete values, the optimization is preliminarily calculated step by step for the groove number from N = 6 to N = 16. Then, the groove numbers, which gives the highest dynamic stiffness, is then selected as the optimum groove number. We had confirmed that the bearing characteristic, which is especially the dynamic stiffness, is decreased if the groove number is larger than N = 16 and less than N = 6 in both cases. In addition, it is difficult to manufacture the grooves on the bearing surface because the width of grooves becomes narrow due to increasing the groove numbers. Therefore, we chose the range of groove numbers from N = 6 to N = 16. This is also to save the computational time for obtaining the optimum solution.

Furthermore, the constraint conditions based on a 2.5 inch HDD thrust bearing are expressed as:

$$g_i(\mathbf{X}) \le 0, \quad (i = 1 \sim 16), \tag{7}$$

Where the details of the constraint functions are defined as in (8)

In (8), besides the constraint function of groove geometries and dimensions from g1 through g 14, the minimum oil lubricating film thickness hr obtained by (2) must be larger than the allowable film thickness to avoid bearing contact and seizures shown as $g15$. Meanwhile, the final constraint function represents the constraint variable of the damping coefficient being kept positive at all times to avoid self-induced vibrations. These values of prescribed constraint values are shown in Table 2:

$$g_1 = r_{s\min} - r_s, \quad g_2 = r_s - r_{s\max}, \quad g_3 = h_{g\min} - h_g,$$
$$g_4 = h_g - h_{g\max}, \quad g_5 = \alpha_{\min} - \alpha, \quad g_6 = \alpha - \alpha_{\max},$$
$$g_7 = \varphi_{1\min} - \varphi_1, \quad g_8 = \varphi_1 - \varphi_{1\max}, \quad g_9 = \varphi_{2\min} - \varphi_2,$$
$$g_{10} = \varphi_2 - \varphi_{2\max}, \quad g_{11} = \varphi_{3\min} - \varphi_3,$$
$$g_{12} = \varphi_3 - \varphi_{3\max}, \quad g_{13} = \varphi_{4\min} - \varphi_4,$$
$$g_{14} = \varphi_4 - \varphi_{4\max}, \quad g_{15} = h_a - h_r, \quad g_{16} = -c.$$

Table 2. Prescribed constraint values

Parameters	**Values**
Groove number N	$6, 7, 8, \ldots, 16$
Minimum seal radius ratio $r_{s\min}$	0.4
Maximum seal radius ratio $r_{s\max}$	0.8
Minimum groove depth $h_{g\min}$ (m)	5.0×10^{-6}
Maximum groove depth $h_{g\max}$ (m)	1.5×10^{-5}
Minimum groove width ratio $\alpha_{\min}$	0.4
Maximum groove width ratio $\alpha_{\max}$	0.9
Minimum geometry parameter $\varphi_{1\sim4\min}$ (deg)	-180
Maximum geometry parameter $\varphi_{1\sim4\max}$ (deg)	180
Allowable film thickness h_a (m)	2.0×10^{-6}, 3.0×10^{-6}, 4.0×10^{-6}

The objective function being set to improve the performance characteristics of the spindle in hydrodynamic bearing is the dynamic stiffness because thrust bearing plays an important role in supporting the total axial direction load of the HDD spindle. The dynamic stiffness, k, of the thrust bearing can be obtained through the following equation:

$$f(\mathbf{X}) = K. \tag{9}$$

The optimization problem for HDD thrust bearing is then formulated as follows:

$$\begin{array}{ll} \max & f(\mathbf{X}) \\ \text{subject to} & g_i(\mathbf{X}) \le 0, \end{array} \quad (i = 1 \sim 16). \tag{10}$$

Vibration Analysis

Once the optimization is being carried out, the vibration analysis is calculated to numerically verify the improvements. The vibration is conducted assuming that the oil film supporting the bearing is a viscously damped one degree free vibration model. The values of the spring and damping coefficient obtained by the optimum design are applied for the vibration response.

The equation of motion is given by

$$m\ddot{x} + c\dot{x} + kx = q(t), \tag{11}$$

where m is the mass of the disk platters, c and k are the damping and spring coefficients obtained by the above mentioned optimum design, respectively, and $q(t)$ is the impulsive external force being applied to the disk platters.

The external impulsive force being applied to the HDD disk platters is assumed as follows:

$$\frac{m\omega_n x(t)}{I} = \left(\frac{e^{-\zeta\omega_n t}}{\sqrt{1-\zeta^2}} \right) \sin \omega_d t, \tag{12}$$

where the natural angular frequency is expressed as $\omega n = \sqrt{k/m}$; the damped natural angular frequency is expressed as $\omega d = \omega n \sqrt{1} - \zeta 2$ which is angular frequency of oscillatory wave form with viscous damping; the damping ratio and impulse are, respectively, expressed as $\zeta = c\ /2\sqrt{mk}$ and $I = f0\Delta t$.The no dimensional expressions for the gain G and phase angle Φ are, then, calculated using the following equations:

$$G = \frac{1}{\sqrt{\left\{1-\left(\omega_f/\omega_n\right)^2\right\}^2 + 4\zeta^2\left(\omega_f/\omega_n\right)^2}}$$

$$\Phi = -\tan^{-1}\frac{2\zeta\left(\omega_f/\omega_n\right)}{1-\left(\omega_f/\omega_n\right)^2}. \quad (13)$$

CHARACTERISTICS OF OPTIMIZED GROOVES

Groove Geometry and Dimensions of Optimized Bearings

Figures 5(a)–5(c) show the groove geometry and pressure distribution of optimized oil lubricated bearings at 10,000 rpm with allowable film thickness of 2.0 µm, 3.0 µm, and 4.0 µm, respectively.

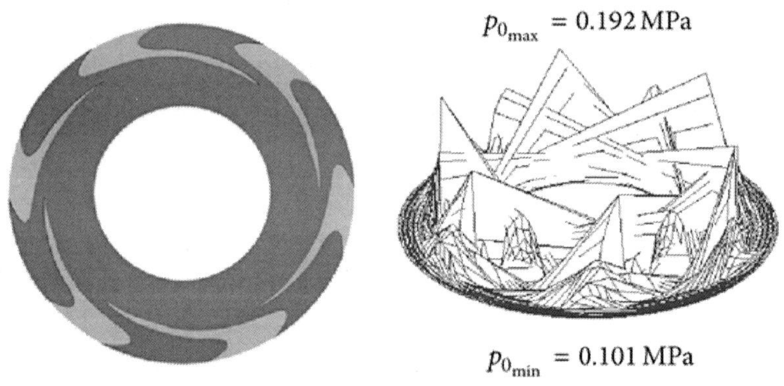

(a)

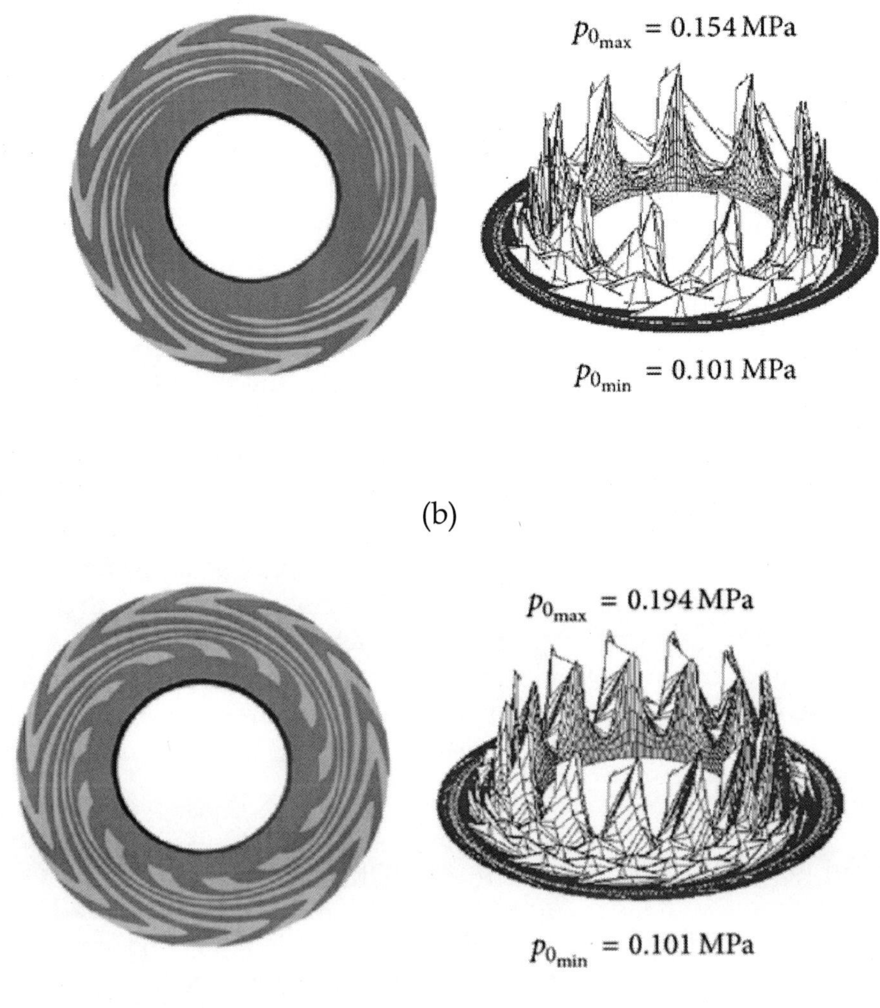

Figure 5: Groove geometry and pressure distribution of optimized oil lubricated bearings. (a) Optimized at 10,000 rpm, *ha*= 2μm, (b) optimized at 10,000 rpm, *ha*= 3μm, (c) optimized at 10,000 rpm, *h a*= 4μm.

As can be seen from the figures, all geometries of the optimized bearings changed from initial spiral groove geometry to a new geometry, a spiral groove with bends in the outer periphery. In this paper, such geometry is called the modified spiral geometry.

Generally, it is widely known that spiral groove geometry has the ability to elevate the film thickness with its pump in effects as the revolution increases due to pressure generation. This explanation is applicable for the inner periphery of the optimized groove geometry. As the bearing rotates, the oil lubricants flows inwards, starting from the outer bends periphery, and flows along the grooves and collides with the seal, hence generating pressure. From all of the pressure distribution figures, the point where the groove and seal meet has the highest point of pressure generated.

However, the phenomenon is different from the outer periphery for opposite spiral groove geometries. Instead of generating the pressure and increasing the film thickness, the oil lubricant flows away from the outer bends towards the most outer radius of the bearing surface. Hence, as can be seen from the pressure distribution figures, there are no pressure peaks generated in the vicinity of the most outer regions of the bearing surface. This is because the pressure in the outer periphery of the bearing is neutralized to the atmospheric pressure. The pressure equivalent to the atmospheric pressure caused by the outer geometry bends will act as a force to pull down the bearing and lower the film thickness. With the combination of spiral geometry in the inner periphery and the opposite spiral bends in the outer periphery, the film thickness decreased, thus maximizing the dynamic stiffness of the spindle.

Even though all groove geometries basically possess the modified spiral geometry, Figure 5(a) shows that the region of spiral groove of optimized with ha= 2.0μm is smaller than those of optimized with ha= 3.0μm and ha =4.0 μm. The geometry is concentrated in the outer regions only, since the seal radius ratio turned to the maximum constraint value being set in this paper, which is 0.8. The oil pressure distribution on the right hand side in the same figure shows pointed sharp pressure distribution. These peaks are the pressure where the groove part and seal part meet, a result of collisions between oil lubricants flow and seal.

For the groove geometries with allowable film thickness of ha= 3.0 μm and ha= 4.0 μm shown in Figures 5(b) and 5(c), the inner part of the bearing possesses a larger portion of spiral groove geometries compared to that optimized with allowable film thickness of ha= 2.0 μm. Even though the inner division creates comparatively smaller pressure peaks, this time the pressure is well distributed throughout the bearing surface.

If we focus on the inner periphery of the total geometry closely, that is, the end part of inner spiral just before the seal, we can see that another bend with the opposite direction of the outer bends started to appear. These inner bends however are smaller than the outer ones. This can be explained by the fact that in order to maintain the prescribed allowable film thickness, it is essential to increase the pump in effect of the inner spiral. It is insufficient with only the inner spiral geometry and outer bends. Therefore, another bend in the inner periphery, just beside the seal, appears to increase the film thickness of the bearing. The inner bends can be seen clearly for the case of allowable film thickness of *ha*= 1.0 µm. The combination of inner bends, outer bends, and inner spiral geometry helps to optimally balance the pressure and improve the dynamic stiffness.

The details of groove geometry and dimension of the optimized bearings from Figure 5 are shown in Table 3. As can be seen from the table, the values of geometry parameters changed from initial to new groove geometry with slightly different angles of through depending on the allowable film thickness. However, in general all the groove geometry changes from spiral groove geometry to spiral with bends in the outer periphery.

Table3- Design variables of optimized bearings at 10,000 rpm

Design variables	Optimized $h_a = 2.0\ \mu m$	Optimized $h_a = 3.0\ \mu m$	Optimized $h_a = 4.0\ \mu m$
Angle φ_1 (deg)	7.94	4.46	2.26
Angle φ_2 (deg)	−20.7	−11.2	−4.70
Angle φ_3 (deg)	79.9	68.1	61.3
Angle φ_4 (deg)	17.5	26.4	32.3
Groove number N	6	11	13
Seal ratio r_s	0.80	0.69	0.63
Groove depth h_g (m)	5.0×10^{-6}	7.5×10^{-6}	8.8×10^{-6}
Groove width ratio α	0.4	0.45	0.48

The other design parameters of groove dimensions such as seal radius ratio, groove depth, and groove width ratio also change. The seal radius ratio shows that the lower the allowable film thickness is being set, the more the groove geometry surface will be occupied with the seal area. This eventually made the groove geometry to be concentrated at the outer region of the total bearing surface. On the other hand, for the groove depth and groove width ratio values, results show that the lower the allowable film thickness is being set, the smaller those values turned into.

Static and Dynamic Characteristics of Optimized Bearings

Figures 6(a), 6(b), and 6(c) show the relations between the rotational speed and the optimized values of dynamic stiffness, *K*, minimum oil lubricating film thickness, *h r*, and friction torque, *Tr*, respectively. The solid lines, alternate long and short dashed lines, and short dashed lines with dots represent the optimized bearings for *ha*= 2.0 μm, *ha*=3.0 μm, and *ha*= 4.0 μm, respectively. The dots represent the characteristics values for each optimized bearing at that specific rotational speed. The short dashed lines without dots represent the initial spiral bearings.

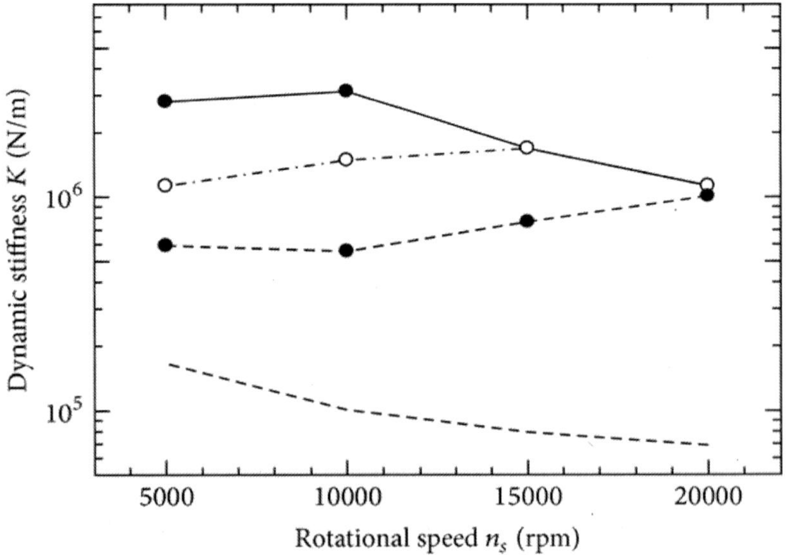

a

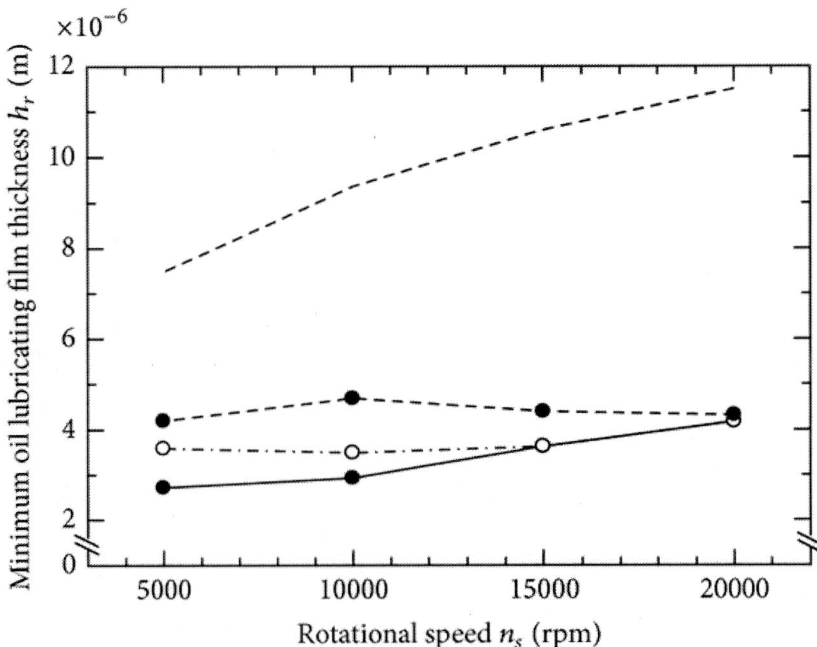
×10^−6
Minimum oil lubricating film thickness h_r (m)
12
10
8
6
4
2
0
5000
10000
15000
20000
Rotational speed n_s (rpm)

b

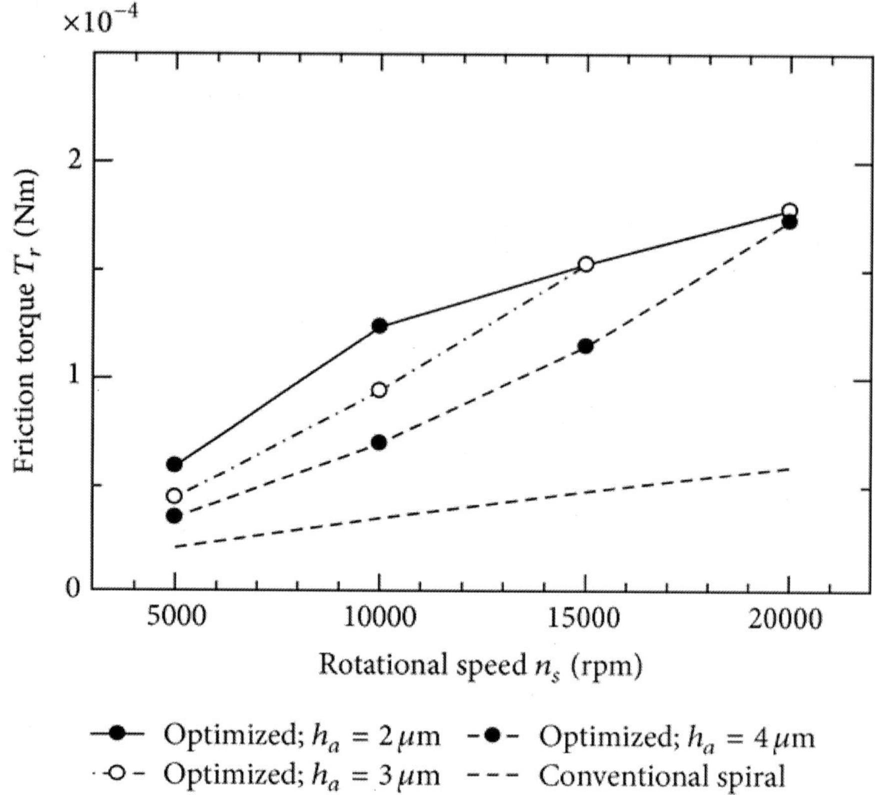

c

Figure 6: Bearing characteristics versus rotational speed; (a) dynamic stiffness *K*, (b) minimum oil lubricating film thickness *hr*, (c) friction torque *Tr*.

The objective function for this study is the improvement of the HDD spindle dynamic stiffness. As can be seen from Figure 6(a), with the proposed optimization method, the dynamic stiffness values are improved compared to the values of conventional spiral bearings. The oil lubricated bearing for conventional spiral showed a trend of decreasing with the increase of rotational speed. However, optimized bearings still show an improvement where different rotational speed gives different stiffness characteristics but with all of them being improved and maintained compared to the conventional ones.

Figure 6(b) shows the relation of minimum oil lubricating film thickness for all optimized results with rotational speed. The minimum oil lubricating film thickness of conventional spiral shows an increase with rotational speed. However, compared to the initial spiral groove, the geometry of optimized bearings has a much lower film thickness. This means that the outer bends of optimized bearings have the ability to pull down the bearings towards bearing base. This is what contributes to the increase of the dynamic stiffness's of the spindle. Moreover, the results obtained are attributable to the groove numbers as shown in Figure 5. Generally, the load carrying capacity is increased with an increase in the groove numbers. Therefore, the oil lubricating film thickness is increased with an increase in the groove numbers because we set constant load (N) in this optimization. On the other hand, the dynamic stiffness is increased with a decrease in the film thickness as shown in Figure 6(a). Therefore, in this optimization, the minimum oil lubricating film thickness comes close to the allowable film thickness to drastically improve the dynamic stiffness changing the groove numbers of ,N= 11, 13 under the allowable film thickness of *ha*= 1.0 μm, *ha*= 3.0 μm, *ha*= 4.0 μm.

For the friction torque values shown in Figure 6(c), all optimized bearings show slightly larger friction torque values compared to initial spiral groove. However, the increase of friction torque is small if we were to compare the dynamic stiffness improvement obtained using the proposed hybrid method.

To confirm the effectiveness of the optimum design presented here, the values of design variables and objective function for the optimized bearing with μm are compared to those for randomly chosen design variables rotating at 10,000 rpm in Table 4. As shown in the table, the objective function of the optimized bearing shows maximum value compared with other values. Furthermore, we confirmed the values of the optimized bearing in the cases of changing allowable film thickness μm, *ha*= 4.0 μm to be the largest using the same method.

Table 4: Comparison of the optimized solution and random solutions under allowable film thickness of *ha*= 3.0 × 10−6 (m).

Case	Obj. $f(X)$ (N/m)	φ_1 (deg)	φ_2 (deg)	φ_3 (deg)	φ_4 (deg)	N	r_s	h_g (m)	α
Optimum solution	1.89×10^6	4.46	−11.2	68.1	26.4	11	0.69	7.50×10^{-6}	0.45
Random 1	2.91×10^5	−60.0	−105.7	−31.4	82.8	6	0.79	1.47×10^{-6}	0.87
Random 2	2.04×10^5	−54.3	−37.1	159.9	162.8	9	0.44	7.90×10^{-6}	0.79
Random 3	4.05×10^5	−131.4	−42.8	−125.7	57.1	11	0.4	1.20×10^{-5}	0.66
Random 4	8.90×10^5	17.1	−5.7	−128.5	88.5	7	0.65	1.08×10^{-5}	0.49
Random 5	1.02×10^6	−168.5	142.8	−97.1	−25.7	10	0.44	1.09×10^{-5}	0.73
Random 6	4.25×10^5	20.0	17.1	−119.9	−105.7	11	0.68	1.09×10^{-5}	0.5
Random 7	1.44×10^6	40.0	−60.0	74.2	−91.4	6	0.58	8.10×10^{-6}	0.58
Random 8	4.85×10^5	77.1	62.8	−117.1	−14.3	12	0.64	1.25×10^{-5}	0.72
Random 9	3.46×10^5	102.8	−42.8	−165.6	−122.8	6	0.71	1.21×10^{-5}	0.41
Random 10	3.54×10^5	131.4	142.8	−142.8	−157.1	8	0.45	1.30×10^{-5}	0.85

Results of Vibration Analysis

Figures 7(a)–7(c) show the comparison of vibration characteristics for optimized bearings with *ha*= 2.0 μm, *ha*= 3.0 μm, *ha*= 4.0 μm, and conventional spiral bearing rotating at 10,000 rpm.

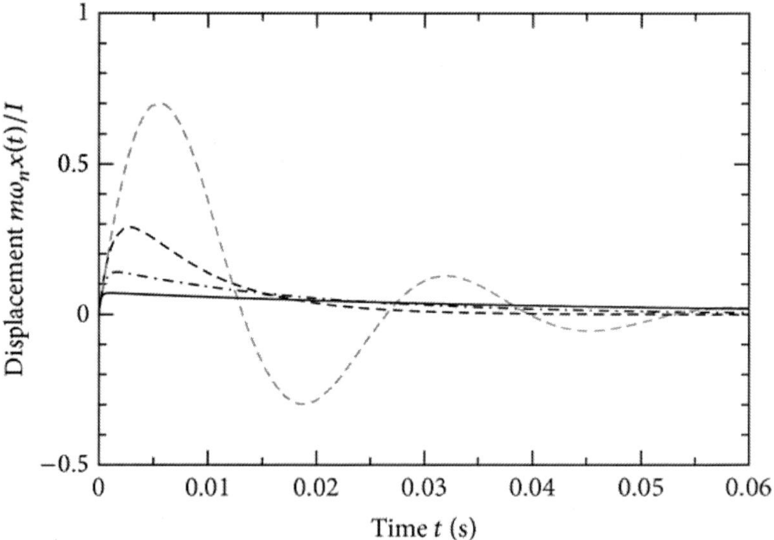

a

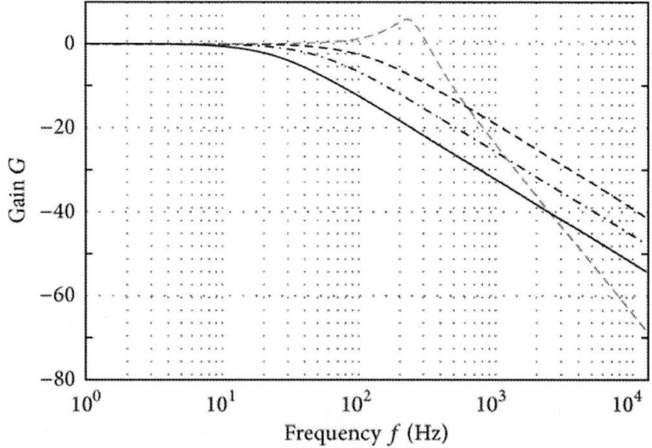

b

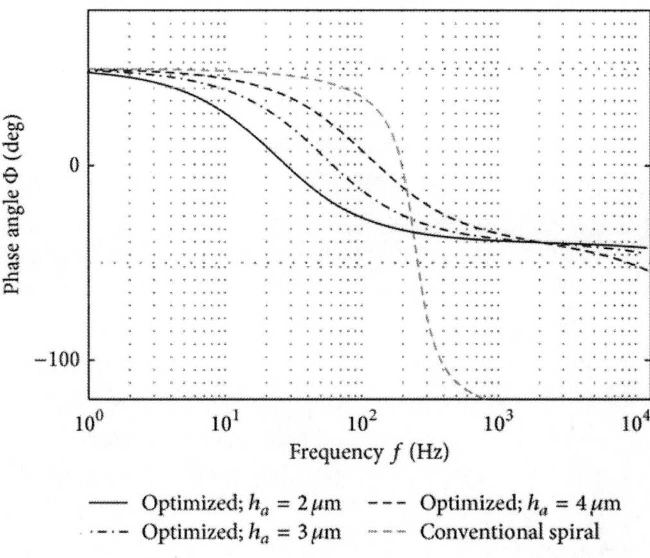

(c)

Figure 7: Vibration characteristics of optimized at 10,000 rpm bearings and conventional spiral bearing; (a) displacement with impulsive force, (b) gain versus frequency, and (c) phase angle versus frequency.

Figure 7(a) shows the response of bearings towards impulsive force. The magnitude of the amplitude for impulsive force shows the highest value for conventional spiral groove bearings where a second order system wave response of an under damped oscillatory response can be seen. On the other hand, the responses towards impulsive force for the rest of the bearings show smaller displacement values where a first order system of an over damped wave form can be seen. This confirms that the optimized bearings have higher stability than the conventional spiral ones.

Figures 7(b) and 7(c) show the frequency response for the bearings. The optimized bearings have a smoother line of gain and phase angle. Gain shows that optimized 2.0 µm have the highest damping followed by 3.0 µm, 4.0 µm, and conventional spiral, respectively. This proves that the optimized bearings have better system responses.

CONCLUSION

In this paper, the optimum design method for 2.5 inch HDD was carried out to improve the dynamic stiffness of a thrust bearing spindle. The results can be concluded as follows.(1)Bearing characteristics using the proposed hybrid method showed that, compared with conventional spiral groove bearings, improvements of dynamic stiffness's can be obtained when a new geometry of modified spiral groove geometry is introduced.(2)To maximize the dynamic stiffness is to set the maximum seal radius ratio, that is, by concentrating the geometry at the most outer periphery of the bearing.(3)Vibration analysis of optimized bearings showed better characteristics than the initial spiral bearings. Therefore, optimized bearings are expected to improve vibration characteristic of HDD

FUTURE WORK

In this paper, we confirmed that dynamic stiffness of thrust bearing for 2.5 inch HDD was drastically improved numerically using the proposed hybrid method. However, it is necessary to verify the calculated data by experiments. We have prepared the experiments to verify the calculated data by using original experimental test rigs as shown in Figure 8. Figures 8(a) and 8(b) show the impulse and vibration experimental test rigs, respectively. The response of bearings towards impulse force, which can be applied using see-saw mechanism as shown in Figure 8(a), can be measured. On the other hand, it is very difficult to measure the spring and damping coefficients of spindle motor for 2.5 inch HDD because the displacement of the bearing is very small. We proposed the identification method [11] of oil film coefficients of bearings for HDD spindle using the response waveform obtained by vibration experiment using the test rig as shown in Figure 8(b). As mentioned above, therefore, we plan to submit the further study including these contents

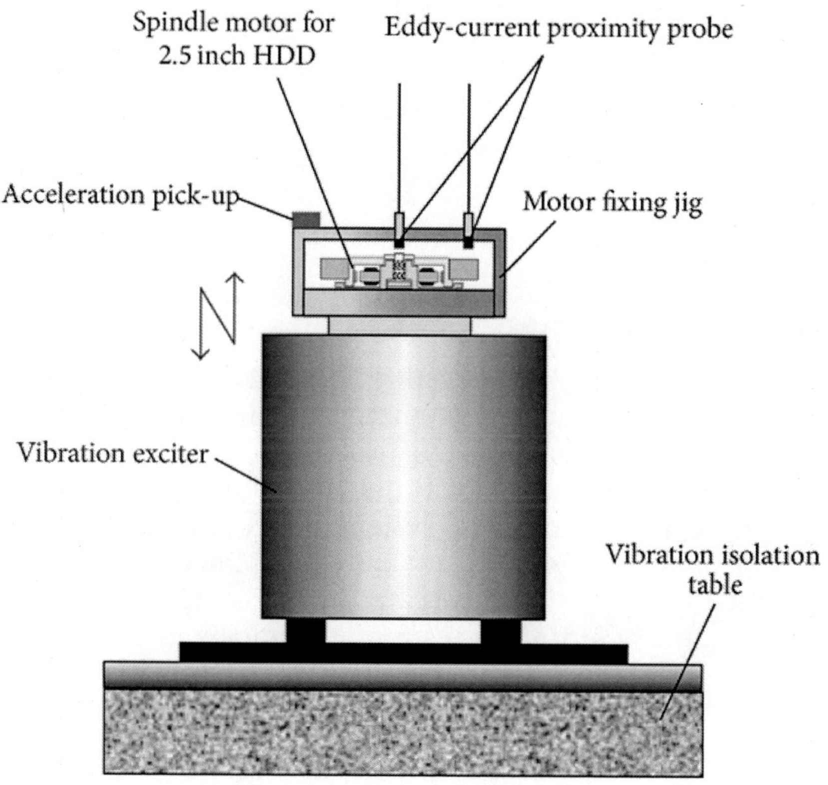

(a) Vibration test rig

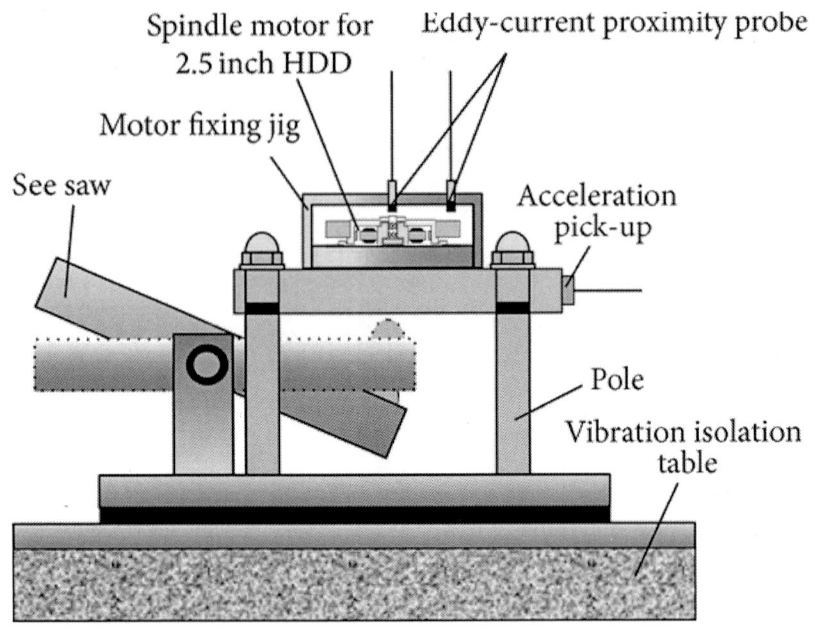

(b) Impulse test rig

APPENDIX

When optimizing the groove geometry, it is necessary to perform a sequential analysis of characteristics of a bearing with a groove geometry modified in succession by the finite difference method, but because of the polar coordinate system, direct processing is rather different to realize. Therefore, an analysis of the bearing stiffness is performed first by using a boundary-fitted coordinate system as shown in Figure 9 [12, 13] and transforming complex groove geometry into a simple fanlike geometry. A boundary transformation function used for the transformation is given as

$$\xi = r$$

$$\eta = \theta - \left[\frac{\theta''(r_i)}{6\Delta r}(r_{i+1} - r)^3 + \frac{\theta''(r_{i+1})}{6\Delta r}(r - r_i)^3 + \left(\theta(r_i) - \frac{\theta''(r_i)\,\Delta r^2}{6} \right) \frac{r_{i+1} - r}{\Delta r} + \left(\theta(r_{i+1}) - \frac{\theta''(r_{i+1})\,\Delta r^2}{6} \right) \frac{r - r_i}{\Delta r} \right]. \quad (A.1)$$

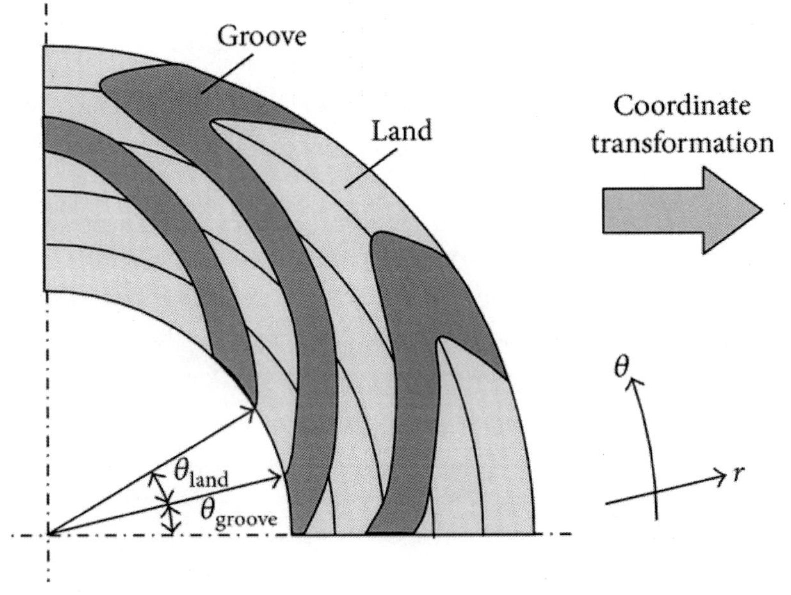

(a)

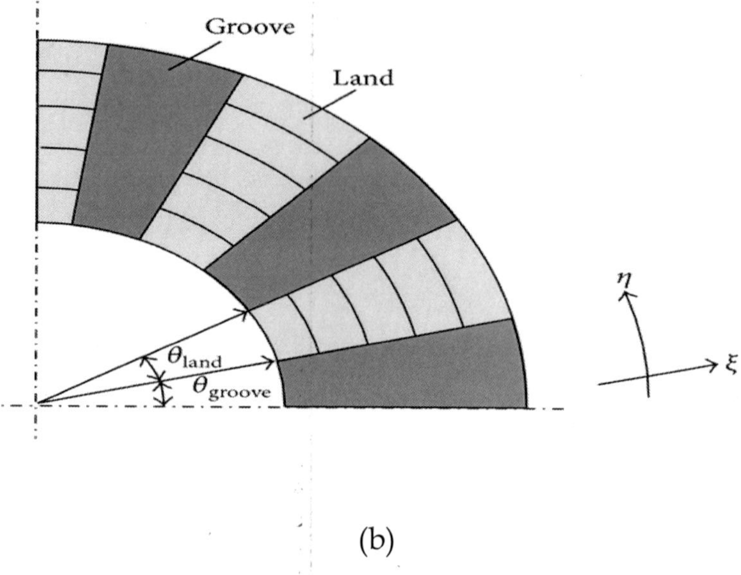

(b)

Figure 9: Bearing geometry transformation based on the boundary-fitted coordinate system; (a) Original bearing geometry, (b) Transformed bearing geometry.

The following Reynolds equivalent equation can be obtained from the equilibrium between the mass flow rates of oil in flowing into and outflowing from the control volume due to the shaft rotation and the squeezing motion:

$$Q_{2I}^{\xi} + Q_{1III}^{\xi} - Q_{2II}^{\xi} - Q_{1IV}^{\xi} + Q_{2I}^{\eta} + Q_{1II}^{\eta} - Q_{2III}^{\eta} - Q_{1IV}^{\eta} = Q^{\Gamma}, \tag{A.2}$$

Where $Q\xi$, $Q\eta$ indicate the mass flow rates, across the boundary of ξ = const. and across the boundary of η = const. as shown in Figure 10. On the other hand, $Q\Gamma$ indicates the mass flow rate due to squeezing motion inside the control volume.

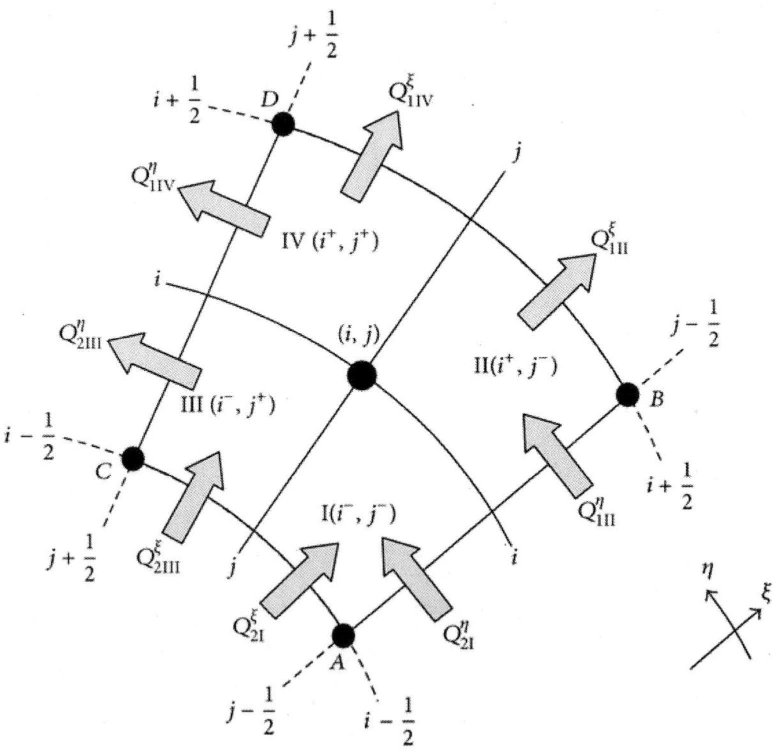

Figure 10: Definition of control volume.

$Q\xi$, $Q\eta$, and $Q\Gamma$ are expressed, respectively, as follows

$$Q^{\xi} = \int_{\eta_1}^{\eta_2} \rho\left(-A\frac{\partial p}{\partial \xi} + B\frac{\partial p}{\partial \eta} + D + E\right) d\eta \tag{A.3a}$$

$$Q^{\eta} = \int_{\xi_1}^{\xi_2} \rho\left(B\frac{\partial p}{\partial \xi} - C\frac{\partial p}{\partial \eta} + F + G\right) d\xi \tag{A.3b}$$

$$Q^{\Gamma} = \int_{\xi_1}^{\xi_2} \int_{\eta_1}^{\eta_2} \frac{\partial(\rho h)}{\partial t} |J|\, d\eta\, d\xi, \tag{A.3c}$$

where, in that case,

$$A = a\frac{h^3}{12\mu J}, \quad B = b\frac{h^3}{12\mu J}, \quad C = c\frac{h^3}{12\mu J},$$

$$D = -\frac{r\omega_s h}{2} r_\eta, \quad E = \frac{\rho r\omega_s^2 h^3}{40\mu} r\theta_\eta, \quad F = \frac{r\omega_s h}{2} r_\xi,$$

$$G = \frac{\rho r\omega_s^2 h^3}{40\mu} r\theta_\xi, \quad a = (r\theta_\xi)^2 + r_\eta^2, \tag{A.4}$$

$$b = (r\theta_\xi)(r\theta_\eta) + r_\xi r_2, \quad c = (r\theta_\xi)^2 + r_\xi^2,$$

$$J = r_\xi(r\theta_\eta) - r_\eta(r\theta_\xi), \quad r_\xi = \frac{\partial r}{\partial \xi}, \quad r_\eta = \frac{\partial r}{\partial \eta},$$

$$r\theta_\xi = r\frac{\partial \theta}{\partial \xi}, \quad r\theta_\eta = r\frac{\partial \theta}{\partial \eta}.$$

Then, in (A.2), subscripts 1 and 2 and I–IV indicate the domains in the control volume.

Assuming that variations of the bearing clearance are microscopic, the minimum oil lubricating film thickness h and pressure p can be expressed by the following equation:

$$h = h_0 + \varepsilon e^{j\omega_f t}$$
$$p = p_0 + \varepsilon p_t e^{j\omega_f t}. \tag{A.5}$$

In the equations stated above, ε indicates the amplitude of small variations of the oil lubricating film thickness and *p*0 and *pt* express a static component and a dynamic component, respectively. The substitution of (A.5) into (A.2) and the negligence of terms of small magnitude of ε of above second order allow for the introduction of two equations for terms ε of orders 0 and 1 as follows:

$$F_0(p_0) = Q^{\xi}_{2I0} + Q^{\xi}_{1III0} - Q^{\xi}_{2II0} - Q^{\xi}_{1IV0} + Q^{\eta}_{2I0} + Q^{\eta}_{1II0} - Q^{\eta}_{2III0} - Q^{\eta}_{1IV0} = 0, \quad \text{(A.6a)}$$

$$F_t(p_t, p_0) = Q^{\xi}_{2It} + Q^{\xi}_{1IIIt} - Q^{\xi}_{2IIt} - Q^{\xi}_{1IVt} + Q^{\eta}_{2It} + Q^{\eta}_{1IIt} - Q^{\eta}_{2IIIt} - Q^{\eta}_{1IVt} - Q^{\Gamma}_{t} = 0, \quad \text{(A.6b)}$$

Where subscript 0 indicates a static component of the mass flow rate determined from (A.3a), (A.3b), and (A.3c), and subscript t similarly indicates a dynamic component. Solving Equations (A.6a) and (A.6b) in turn by the Newton-Rap son iteration method, the static and dynamic components, $p0$ and pt, are obtained

NOMENCLATURE

$b1$: Groove width (m)

$b2$: Land width (m)

c: Damping coefficient of oil lubricated film(N·s/m)f(X): Objective functiongi(X): Constraint functions; (i = 1 ~ 16)q(t): Impulse external force (N)G: Gain$h$

a: Allowable film thickness (m)h

g: Groove depth (m)h

r: Minimum oil lubricating film thickness (gap between shaft and lower groove surface) (m)

k: Spring coefficients of oil lubricated film (N/m)

K: Dynamic stiffness of oil lubricated film (N/m)

m: Mass of the rotor (kg)

N: Groove number n

s: Rotational speed of spindle (rpm)

p: Oil lubricated film pressure (Pa)

a: Atmospheric pressure (Pa) p

0: Static component of oil lubricated film pressure (Pa)

t: Dynamic component of oil lubricated film pressure (Pa/m)

r: Coordinate of radius direction (m) Rout: Bearing outer radius (m)

*R*in: Bearing inner radius(m) *Rs*: Seal radius(m)*rs*: Seal ratio;= (*Rs*/*R*out)*t*: Times (s)

Tr: Friction torque on bearing surface (N·m) *W*: Load-carrying capacity (N) X: Vector of design variables Δr: Equipartition space of *r* (m).

REFERENCES

1. G. H. Jang, S. H. Lee, and H. W. Kim, "Finite element analysis of the coupled journal and thrust bearing in a computer hard disk drive," Journal of Tribology, vol. 128, no. 2, pp. 335–340, 2006. View at Publisher · View at Google Scholar · View at Scopus
2. K. Y. Park and G. H. Jang, "Dynamics of a hard disk drive spindle system due to its structural design variables and the design variables of fluid dynamic bearings," IEEE Transactions on Magnetics, vol. 45, no. 11, pp. 5135–5140, 2009. View at Publisher · View at Google Scholar · View at Scopus
3. T. Kobayashi and H. Yabe, "Numerical analysis of a coupled porous journal and thrust bearing system," Journal of Tribology, vol. 127, no. 1, pp. 120–129, 2005. View at Publisher · View at Google Scholar · View at Scopus
4. J. Zhu and K. Ono, "A comparison study on the performance of four types of oil lubricated hydrodynamic thrust bearings for hard disk spindles," Journal of Tribology, vol. 121, no. 1, pp. 114–120, 1999. View at Scopus
5. T. Asada, H. Saitou, and D. Itou, "Design of hydrodynamic bearing for miniature hard disk drives," IEEE Transactions on Magnetics, vol. 43, no. 9, pp. 3721–3726, 2007. View at Publisher · View at Google Scholar · View at Scopus
6. G. H. Jang and C. I. Lee, "Development of an HDD spindle motor with increased stiffness and damping coefficients by utilizing a stationary permanent magnet," IEEE Transactions on Magnetics, vol. 43, no. 6, pp. 2570–2572, 2007. View at Publisher · View at Google Scholar · View at Scopus
7. Q. Zhang, S. Chen, S. H. Winoto, and E.-H. Ong, "Design of high-speed magnetic fluid bearing spindle motor," IEEE Transactions on Magnetics, vol. 37, no. 4, pp. 2647–2650, 2001. View at Publisher · View at Google Scholar · View at Scopus
8. Y. Arakawa, S. Ikeda, T. Hirayama, T. Matsuoka, and N. Hishida, "Hydrodynamic bearing with non-uniform spiral grooves for high-

speed HDD spindle," in Proceedings of the IEEE 13th International Symposium on Consumer Electronics (ISCE '09), pp. 511–512, Kyoto, Japan, May 2009. View at Publisher · View at Google Scholar · View at Scopus

9. H. Hashimoto and M. Ochiai, "Optimization of groove geometry for thrust air bearing to maximize bearing stiffness," Journal of Tribology, vol. 130, no. 3, Article ID 031101, 2008. View at Publisher · View at Google Scholar · View at Scopus

10. M. D. Ibrahim, T. Namba, M. Ochiai, and H. Hashimoto, "Optimum design of thrust air bearing for hard disk drive spindle motor," Journal of Advanced Mechanical Design, Systems and Manufacturing, vol. 4, no. 1, pp. 70–81, 2010. View at Publisher · View at Google Scholar · View at Scopus

11. M. Ochiai, Y. Sunami, and H. Hashimoto, "Experimental study on dynamic characteristics of fluid film bearing for HDD spindle motor," in Proceedings of the 2nd International Conference on Design Engineering and Science, pp. 217–222, 2010.

12. H. Hashimoto and M. Ochiai, "Theoretical analysis and optimum design of high speed gas film thrust bearings (static and dynamic characteristic analysis with experimental verifications)," Journal of Advanced Mechanical Design, Systems, and Manufacturing JSME, vol. 1, no. 1, pp. 102–112, 2007.

13. N. Kawabata, "Study on generalization of calculations of lubricant flow using boundary fitted coordinate system (part 1, basic equations of DF method and case of incompressible fluids)," Transactions of the Japan Society of Mechanical Engineers C, vol. 53, no. 494, pp. 2155–2160, 1987. View at Scopus

Citations

CHAPTER 1

Ricardo Ugliara Mendes and Katia LucchesiCavalca, "On the Instability Threshold of Journal Bearing Supported Rotors," International Journal of Rotating Machinery, vol. 2014, Article ID.

CHAPTER 2

KaprawiSahim, KadafiIhtisan, DyosSantoso, and RimanSipahutar, "Experimental Study of Darrieus-Savonius Water Turbine with Deflector: Effect of Deflector on the Performance," International Journal of Rotating Machinery, vol. 2014, Article ID 203108, 6 pages, 2014. doi:10.1155/2014/203108

CHAPTER 3

G. Belforte, F. Colombo, T. Raparelli, A. Trivella and V. Viktorov (2013). High Speed Rotors on Gas Bearings: Design and Experimental Characterization, Tribology in Engineering, Dr. HasimPihtili (Ed.), ISBN: 978-953-51-1126-9, InTech, DOI: 10.5772/50795.

CHAPTER 4

M. Khosravyel_Hossaini (2013). Review of the New Combustion Technologies in Modern Gas Turbines, Progress in Gas Turbine Performance, Dr. Ernesto Benini (Ed.), ISBN: 978-953-51-1166-5, InTech, DOI: 10.5772/54403.

CHAPTER 5

Xiao-lu Lu, Kun Zhang, Wen-hui Wang, Shao-ming Wang, and Kang-yao Deng, "Preliminary Experimental Study on Pressure Loss Coefficients of Exhaust Manifold Junction," International Journal of Rotating Machinery, vol. 2014, Article ID 316498, 10 pages, 2014. doi:10.1155/2014/316498

CHAPTER 6

Yannick Bousquet, Xavier Carbonneau, Guillaume Dufour, Nicolas Binder, and Isabelle Trebinjac, "Analysis of the Unsteady Flow Field in a Centrifugal Compressor from Peak Efficiency to Near Stall with Full-Annulus Simulations," International Journal of Rotating Machinery, vol. 2014, Article ID 729629, 11 pages, 2014. doi:10.1155/2014/729629

CHAPTER 7

Man Zhao, Shurong Yu, Chao Li, and Yang Yu, "Effect of Mechanism Error on Input Torque of Scroll Compressor," International Journal of Rotating Machinery, vol. 2013, Article ID 438127, 5 pages, 2013. doi:10.1155/2013/438127

CHAPTER 8

Yuta Sunami, Mohd Danial Ibrahim, and Hiromu Hashimoto, "Optimum Design of Oil Lubricated Thrust Bearing for Hard Disk Drive with High Speed Spindle Motor," International Journal of Rotating Machinery, vol. 2013, Article ID 896148, 11 pages, 2013. doi:10.1155/2013/896148

INDEX